PMM D&T AND AUTOBIOGRAPHY OF THE INVENTOR

Nathan Coppedge

Nathan Coppedge

THE DESIGNS & THEORY

AND

THE AUTOBIOGRAPHY OF THE INVENTOR OF PERPET-UAL MOTION

=BOUND SET=

By Nathan Coppedge

Nathan Coppedge

-INDEX-

Nathan Coppedge

-INTRODUCTION-

You may not know this, but at first this book did not contain an introduction. About that time, I started receiving negative reviews. People were losing patience with the diagrams, which led to a complete obfuscation of the themes of the work.

How could this be? I asked myself! After all, I was one of the few pioneers to develop what seemed to me like a real viable theory of perpetual energy. I won't consider the other prospective inventors. They were marred with over-complexity, an obsession with magnetism, etc. I could produce simple proofs that did away with the theories of other men.

How could it be that the purity of my golden ideas was being rejected by my first critics?

As I have already said, the problem was that my critics didn't think I took myself seriously. They thought I had generated garbage ideas which required a physical model to establish proof. However, this was very far from my intention.

The theories in this book may include some failures. I recognize that not all of my designs are viable. But I also recognized that I, unlike many others, showed some signs of success. Readers, however, should recognize the long tradition of failure which underscores any development in this field. It is the small ink-lings of a new idea, an inspired genius, which we should identify and improve.

This is a book of ideas: both theories and concepts, which was not developed instan-taneously. Consider that time and effort con-tributed as much as the occasional epiphany. Readers should recognize that the knowledge contained in this book does not come automati-cally. Like any serious intellectual field, under-standing it requires subtlety and precision.

Readers should also recognize that this is not a practical guide for building the de-vices. Doubtless countless examples of that kind will spring up if there is, for once, a vi-able example of success that is simple enough to reproduce. This book, on the other hand, comes earlier in the timeline than the practical manuals.

It is, so to speak, an esoteric guide to a much misunderstood field. It provides the groundwork for the genius of over-unity. It provides shortcuts and common themes that have for centuries gone overlooked.

This manual is an attempt to catch up with history. Therefore, you will understand if interpreting it requires a leap of the imagina-tion.

THEORIES

Nathan Coppedge

3 PROOFS

PROOF #1: FRICTION DOES NOT ELIMINATE MOTION WHERE MOTION IS PERMITTED.

PROOF #2: REACTIONS ARE POSSIBLE IN A CIRCLE, AS SHOWN BY DOMINOES. WHEELS CAN TURN!

PROOF #3: DOMINOES CAN CHAIN-REACT USING HIGHER AND HIGHER ALTITUDES. ENERGY CAN BE CREATED!

"VOLITIONAL MECHANICS"

FIRST PRINCIPLE: Continuous Motion:
However unbalanced the parts, there must be a method of continuum, a means such as ramps and differences in force, to allow motion to continue. This is the most central principle, but by itself it does not yield much technicalism.

SECOND PRINCIPLE: Unity
In order to meet criteria of unity in a simple device, all parts must interact via equilibrium, such as over an axle, or through circular motion. If not, then this suggests use and disuse of parts, which is still a function of equilibrium. Use and disuse can also be an effective principle, if there is no change-of-altitude problem.

THIRD PRINCIPLE: Over-Unity
Over-unity is essentially the principle of motion in spite of equilibrium. This can be done through directed energy (continuous slope, as in the Tilt Motor), differential energy (mass-space differences, as in Repeat Lever Type 2), or an unbalanced principle (methods of "cheating," as in the Trough-type devices).

 FOURTH PRINCIPLE: Volitional Energy is represented by the Volit (pictured), representing an effective cycle. Devices tend to be more functional not only with a good principle, but with a large number of moving parts in comparison to fixed and dual-axial parts. To my knowledge this has borne out with the exception of "cheating" principles, in which fixed units may actually have an advantage.

17

Ultimately I decided dual-axialism was a more important disadvantage than fixed parts in the right design. But, instructively, both may be considered bad. Volitional Energy is expressed as:

= Active Units / Dual-Axial Units

FIFTH PRINCIPLE: Volitional Equilibr. (Ve)
Generally, a large number of modules is good when it works, and a large number of channels or branches in the movement is bad. Not all devices use modules, so this value is often 1. It is expressed as:
Ve = (mod. U /
(branches / cycle) / (subcycles / cycle))

SIXTH PRINCIPLE: Volitional Eff. (VE)
Overall efficiency is estimated as an equation between volitional energy and volitional equilibrium. For example, if volitional energy (also called volition) is 2, and equilibrium is 1, the result is 2. If the equilibrium is 4, the result is 0.5. If the equilibrium is 0.5, the result is 4. Higher numbers are better. You will find for most viable principles, the number is hardly ever above 2. '1' represents unity, hypothetically, and the number must be greater than 1 to equal perpetual motion.

$$VE = \frac{\text{ʊ}}{Ve}$$

Three Types of Perpetual Energy

Continuous Mass: Applying weight without motion

Continuous Velocity: Photons that have momentum without mass

Continuous Momentum: Machines that are unbalanced at every point in a cycle

Principles of Volitional Energy

Principle of Volitional Energy: *Mass that can by some means move on its own, and repeat this motion, has a form of momentum.*

Principle of Volitional Equilibrium: *Force within a cycle that does not divide is singular, whether or not it encounters resistance.*

Principle of Volitional Efficiency: *When the division of force is less than the implication of momentum, the function of a device scales to design principles, rather than mere re-enactment of basic laws understood in basic ways.*

There is freedom to create a machine whenever such a machine is possible to build. The process is a posteriori and not a-priori: it is found after the fact. As I will show later on, perpetual motion is a special case of specific relations of laws, laws which do not contradict the most reasonable perspective on conventional physics, in all conventional cases, and certainly would not contradict all unconventional physics, in unconventional cases.

Major Theory Principles

CYCLES CAN RETURN TO THEIR STARTING POINT

Consider a circle of dominoes. The dominoes can return to the starting point. The problem is re-setting the cycle. Similarly, a wheel can be made to rotate in a circle many times, thus repeating the cycle; the problem is creating motion in the first place. These examples prove that cycles can certainly return to their starting points. There is no physical law against recursive movement.

PERPETUAL MOTION DOES NOT REQUIRE ENERGY

Momentum is Possible Without Velocity, Therefore Momentum is Possible Without "Energy": Just as photons have momentum without mass according to the basic rule of Mass * Velocity = Momentum, in certain cases mass that is not moving still exerts force, and since the mass does not move, the force is not lost. However, that force, according to my rules, is understood as momentum without velocity. An example of momentum that exerts "force" without losing energy is the case of a rolling object tethered horizontally upon a slope, with the tether leading horizontally (say, through a split in the sloped surface) beyond the slope, to some other fixed object. It is clear that the tether remains taut if the rolling object is heavy or if the slope is steeply inclined, yet with some allowance for control, there is no reason to believe that the tautness is caused by an actual movement of the rolling weight. For example, consider that any initial movement could be easily repeated so long as slope remains for motion. But there is no necessity that movement occurs in the first place at all. Certainly it is not a requisite for making the cord, rope, or string taut. The point is, however, that a general principle of momentum without velocity is potentially foundational for perpetual motion, because most physicists seem to have been assuming that energy was necessary for perpetual motion. The fogginess of the basic equation in which velocity or momentum replace energy suggests that any one of the three---velocity, mass, or momentum---may be energy in itself. While this does not necessarily make sense in the strictest physics equations, it is a foregone conclusion that each of those three aspects mentioned *does* connote energy in some sense, that is, they could mean potential energy *if* we assume gravity, and make the radical stipulation that gravity is not always consumed. We do not at this stage have to posit the creation of energy---indeed, the creation of energy must be a special case---and

21

it is already by the adoption of a special case that one individual principle, such as velocity, mass, or momentum, could become viable. According to the principles of volitional energy, fractionism implies that only one of a coherent set of principles may be necessary to simulate the entire conjunction, that is, if the other principles are also present at other points. The principles of spatial difference and fractionism underlie this principle, of momentum without velocity.

SYSTEMS CAN GAIN ENERGY

A case is raised of dominoes set up in increasing heights. Some people suspect that this would not work. Taken to a far extreme, they're right. They expect that a domino won't knock down a building for instance. It won't. However, one domino can knock down two dominoes stacked up. Try it. It works. Further, the two dominoes can knock over three dominoes. And the three dominoes can knock down four dominoes. It can keep going ad infinitum so long as it is possible to stack up the dominoes. Some would say that this flies in the face of physics. I would say that it is a fine application of physics. Also, consider as a stipulation that the altitude rule is not what is applying here. Don't be fishy about this issue. Increasing heights does not mean increasing altitudes, inherently. The dominoes could be stacked at decreasing altitudes, and they would still operate. That is enough to prove the point. Or, the first domino could touch a stick that reaches up to the top of a stack that then causes a down-ward chain reaction. How the chain reaction occurs has little to do with altitude, except in relation to changes in the gravitational force. Certainly we wouldn't say that a change in the gravitational force is what causes a chain reaction. Instead we would say that 'inputted' force causes motion (momentum), and in this case it is evident that a small inputted force is sufficient to cause a mass reaction. One of the only principles preventing this from becoming a self-repeating cycle is the abstract principle that the cycle cannot be re-set. It is obvious that the system can gain energy.

Major Theory Applications
*Energy
>Power & Heat
*Mobility
>Linear Transit through Power
>Cyclical Transit through Cont. Motion
*Social Reform
>Energy Independence
>Ecology & Arcology
>Free Money
*Symbolism
>Immortality
>Industry
>Intelligence
>Fun

Minor Theory Applications
*Continuous Motion
*Toys and Novelties
*Frivolity
*Intellectual Concepts

Perpetual Motion Types

Rank of Types and Variations

Escher Machine: ∞

Mod. Trough Leverage: ∞ / 8

Spin Devices: 1, 2, or 1 or 2 ∞
Tilt Motor: 5
Motive Mass Machine: 3
Trough-Leverage: 2
Repeating Lever: 2
Magnet Device: 2
Coquette: 2
Bezel Weight Device: 1.5
Gravity-Buoyancy Device: ~1
Curvilinear-Rail Machine: ~1
Gravity Machines: 1
Fluid Leverage: 1
Apollo Device: 1

Nathan Coppedge

DESIGNS

Nathan Coppedge

EARLY FAILURES

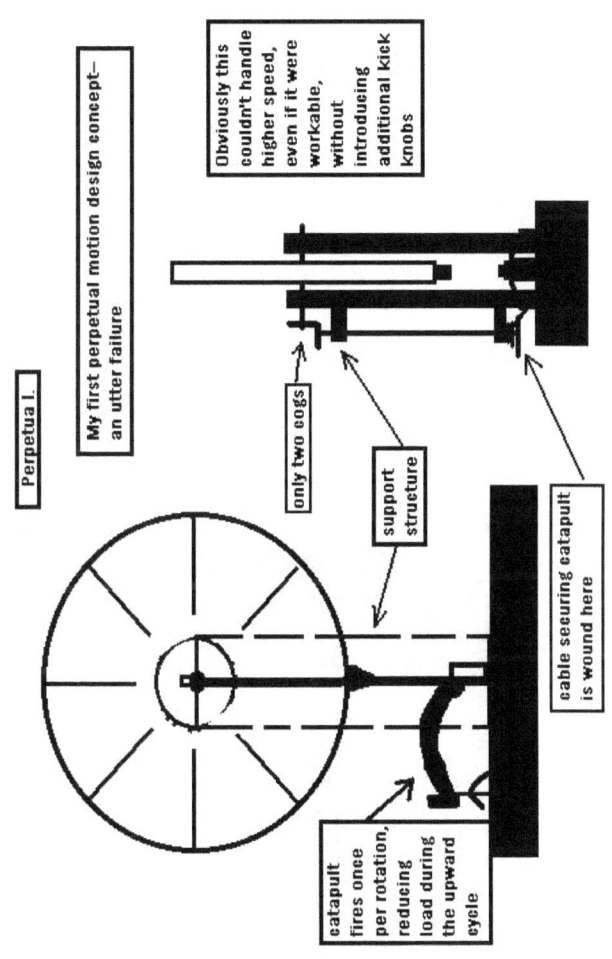

Perpetua I.

My first perpetual motion design concept—an utter failure

Obviously this couldn't handle higher speed, even if it were workable, without introducing additional kick knobs

only two cogs

support structure

cable securing catapult is wound here

catapult fires once per rotation, reducing load during the upward cycle

TYPE 1

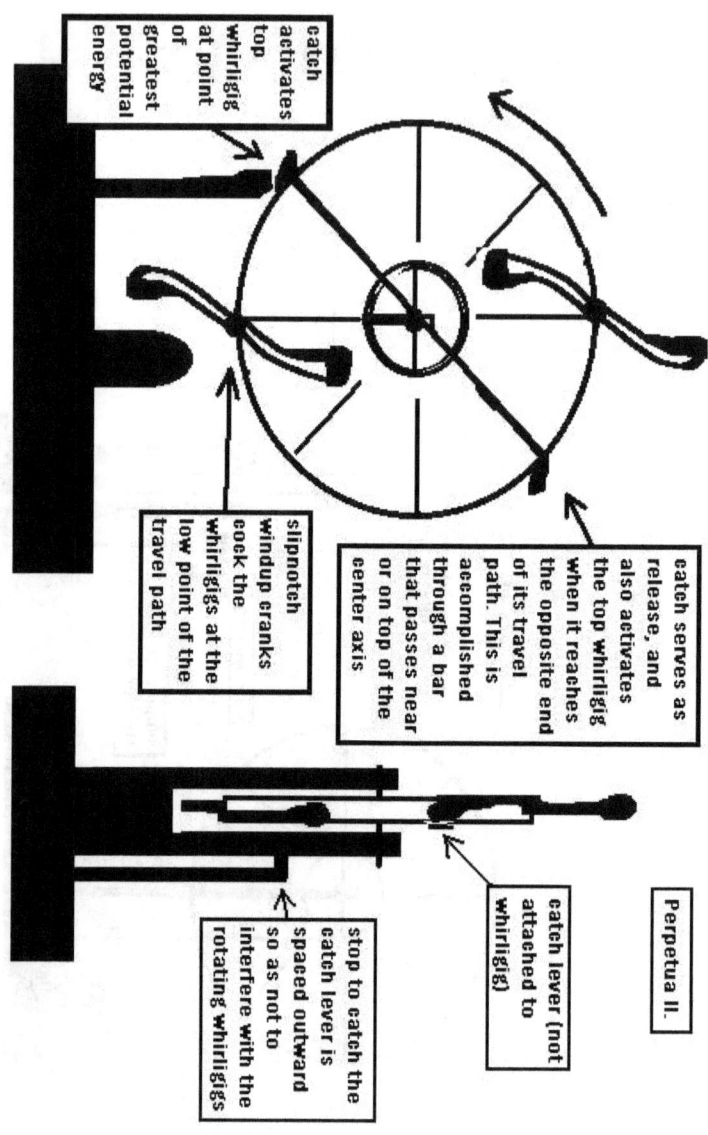

catch activates top whirligig at point of greatest potential energy

slipnotch windup cranks cock the whirligigs at the low point of the travel path

catch serves as release, and also activates the top whirligig when it reaches the opposite end of its travel path. This is accomplished through a bar that passes near or on top of the center axis

Perpetua II.

catch lever (not attached to whirligig)

stop to catch the catch lever is spaced outward so as not to interfere with the rotating whirligigs

EARLY CONCEPT 2

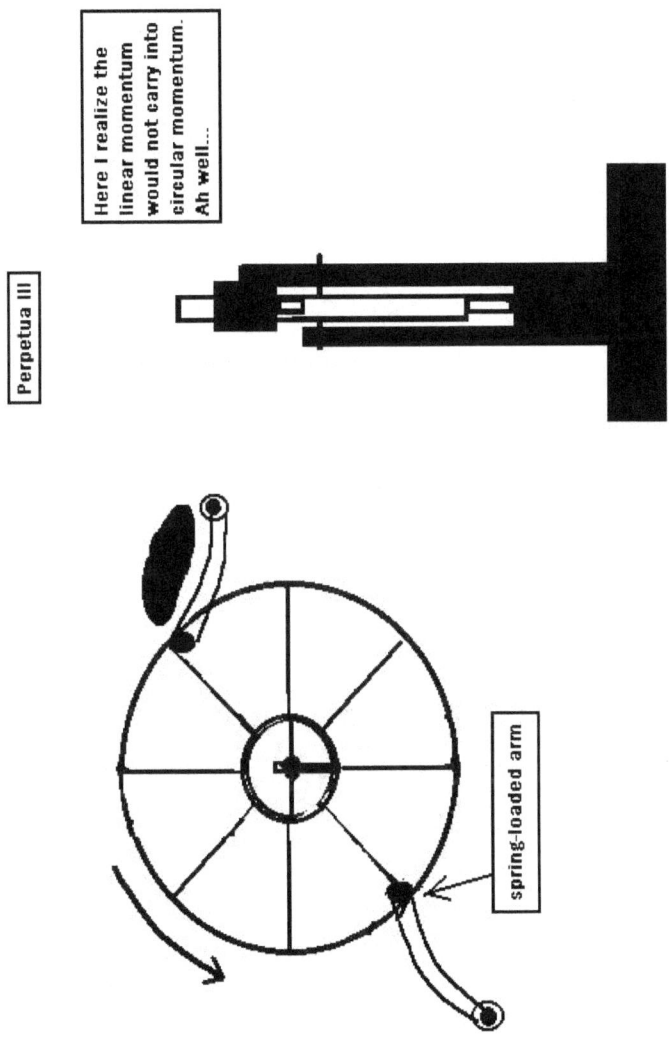

Perpetua III

Here I realize the linear momentum would not carry into circular momentum. Ah well....

spring-loaded arm

EARLY CONCEPT 3

GRAV-BUOY ITERATION 1

Filed Ma

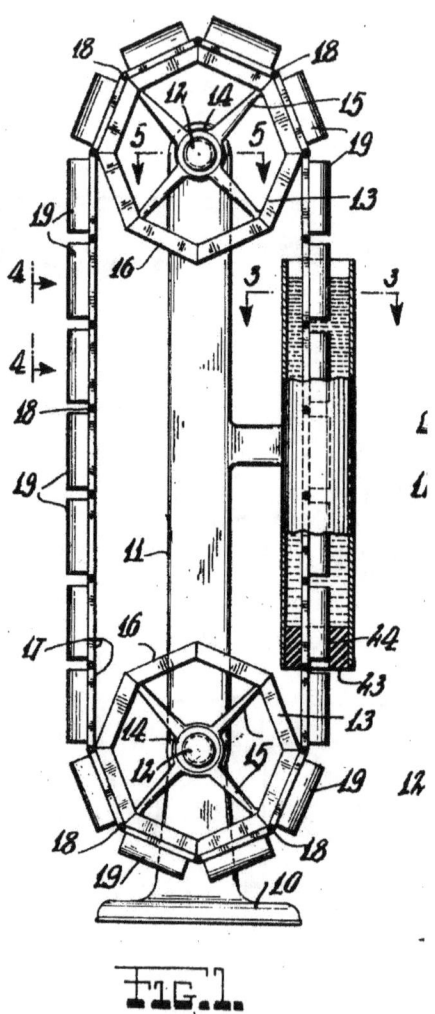

FIG.1.

FRANK TATAY'S DESIGN

GRAV-BUOY ITERATION 2
COMPONENTS

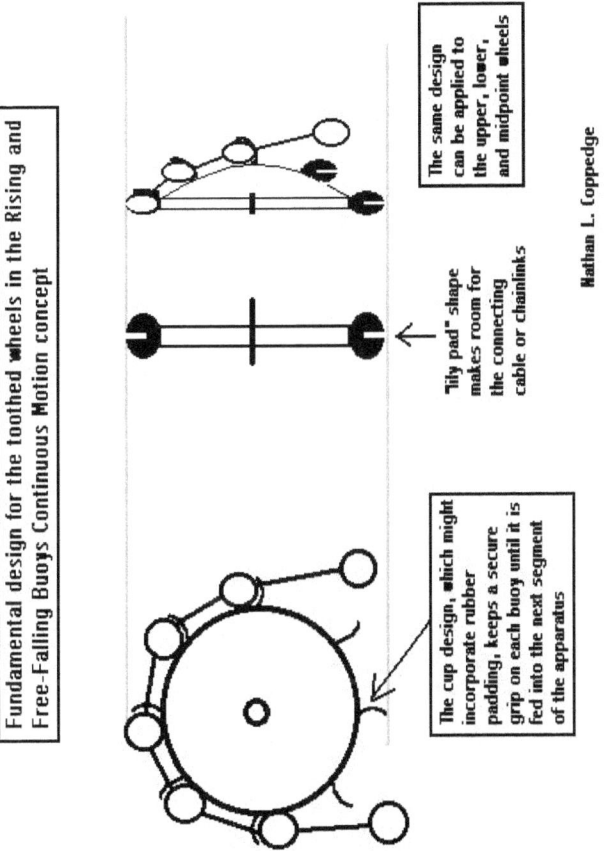

Fundamental design for the toothed wheels in the Rising and Free-Falling Buoys Continuous Motion concept

The same design can be applied to the upper, lower, and midpoint wheels

Nathan L. Coppedge

"lily pad" shape makes room for the connecting cable or chainlinks

The cup design, which might incorporate rubber padding, keeps a secure grip on each buoy until it is fed into the next segment of the apparatus

TOOTHED WHEEL FOR THE GRAV-BUOY 2

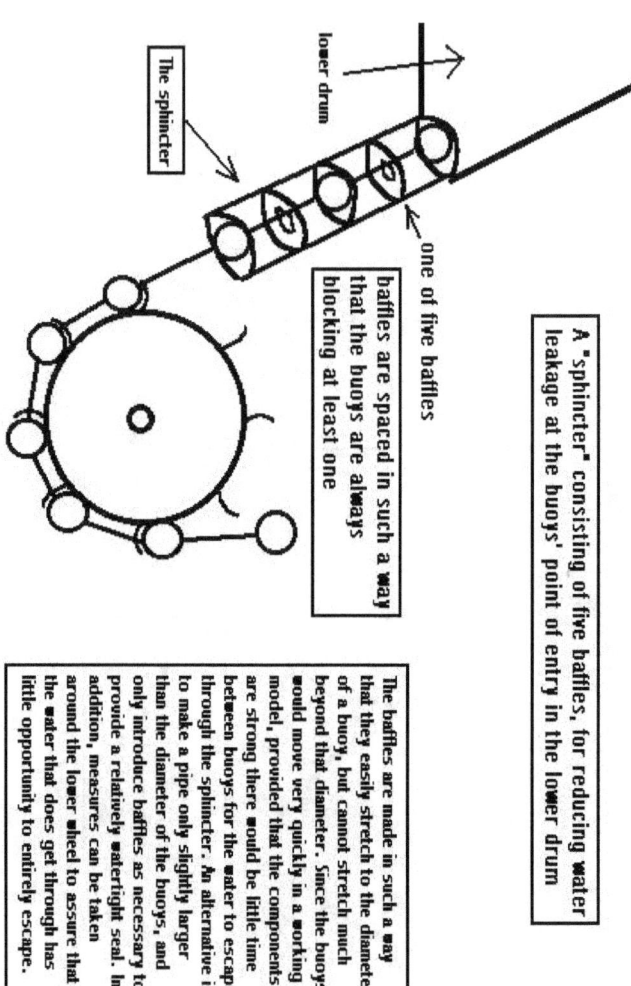

lower drum

The sphincter

one of five baffles

A "sphincter" consisting of five baffles, for reducing water leakage at the buoys' point of entry in the lower drum

baffles are spaced in such a way that the buoys are always blocking at least one

The baffles are made in such a way that they easily stretch to the diameter of a buoy, but cannot stretch much beyond that diameter. Since the buoys would move very quickly in a working model, provided that the components are strong there would be little time between buoys for the water to escape through the sphincter. An alternative is to make a pipe only slightly larger than the diameter of the buoys, and only introduce baffles as necessary to provide a relatively watertight seal. In addition, measures can be taken around the lower wheel to assure that the water that does get through has little opportunity to entirely escape.

SCHINCTER FOR THE GRAV-BUOY 2

GRAVITY-BUOYANCY DEMO

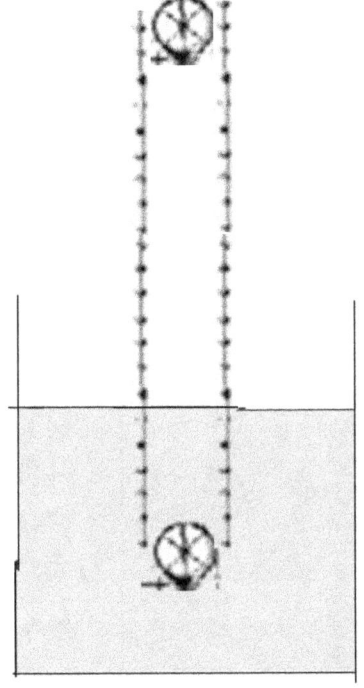

HYPOTHESIS 1

BASIS: FRANK T.

LOWERING THE
WATER LEVEL
IS THE BEST
THEORY AT
FIRST GLANCE
HOWEVER,
THIS PROVES
EQUILIBRIUM

GRAV-BUOY DEMO 1

HYPOTHESIS 2

ALTERING THE ANGLE OF THE DEVICE IS THE EASIEST MOST LIKELY WAY TO EFFECT A CHANGE IN THE APPLICABILITY OF UNIVERSAL LAW

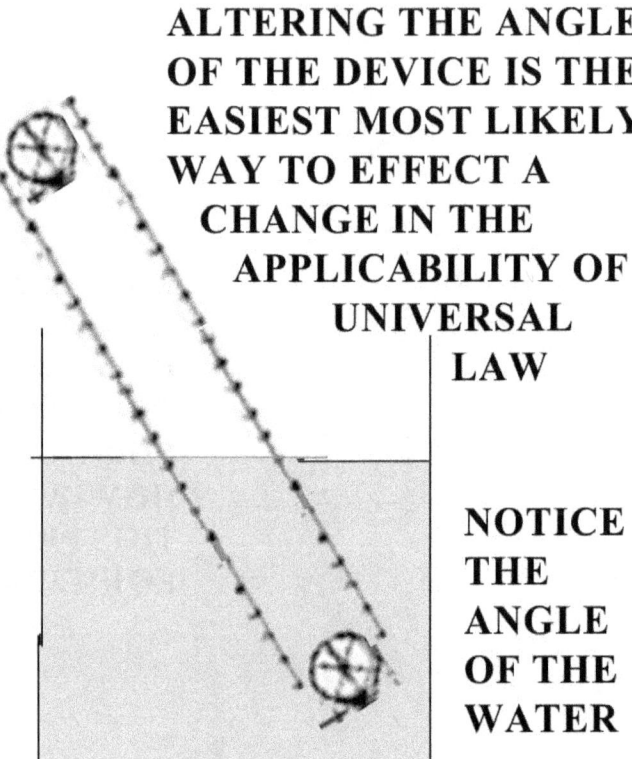

NOTICE THE ANGLE OF THE WATER

GRAV-BUOY DEMO 2

HYPOTHESIS 3

A FURTHER PRINCIPLE WOULD MAKE SOME USE OF AIR PRESSURE HOWEVER, THE POSITION IS INITIALLY CONFOUNDED

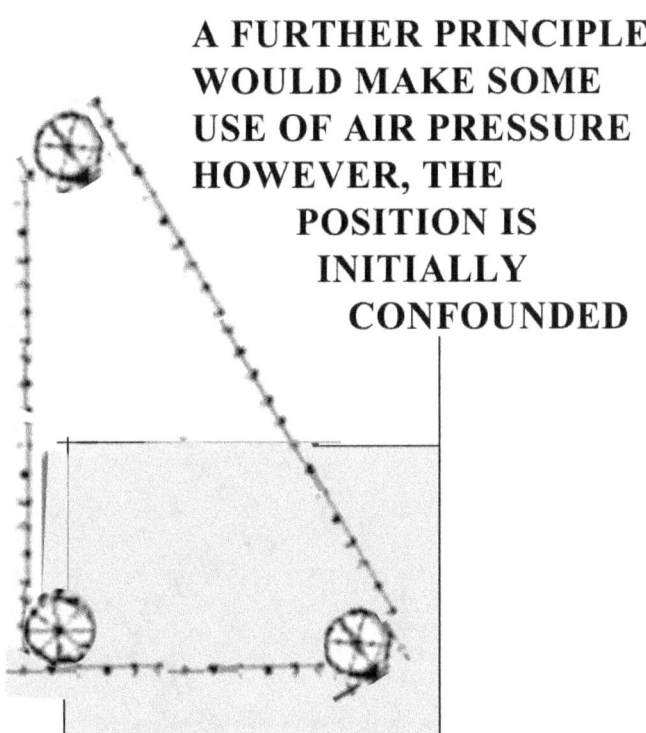

GRAV-BUOY DEMO 3

Nathan Coppedge

HYPOTHESIS 4

SOME SORT OF WATER TANK MIGHT BE RE-INTRODUCED, BUT THE PROBLEM IS WITH A BALANCE BETWEEN WEIGHT AND BUOYANCY, WITH DEVICIVE WATER PRESSURE AS AN INTERMEDIATE

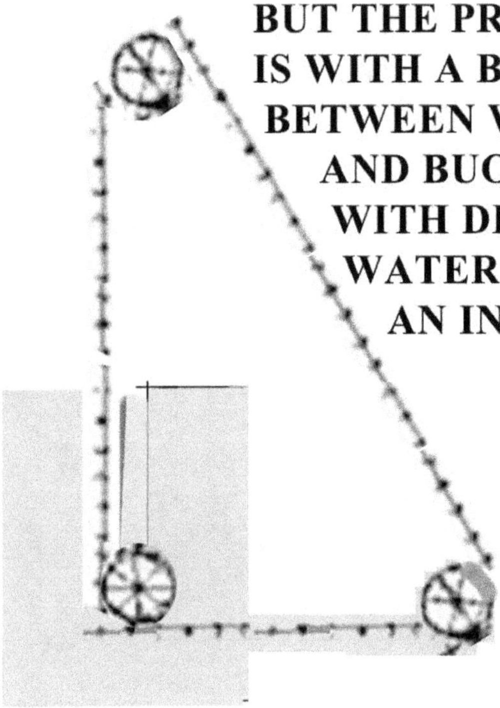

GRAV-BUOY DEMO 4

38

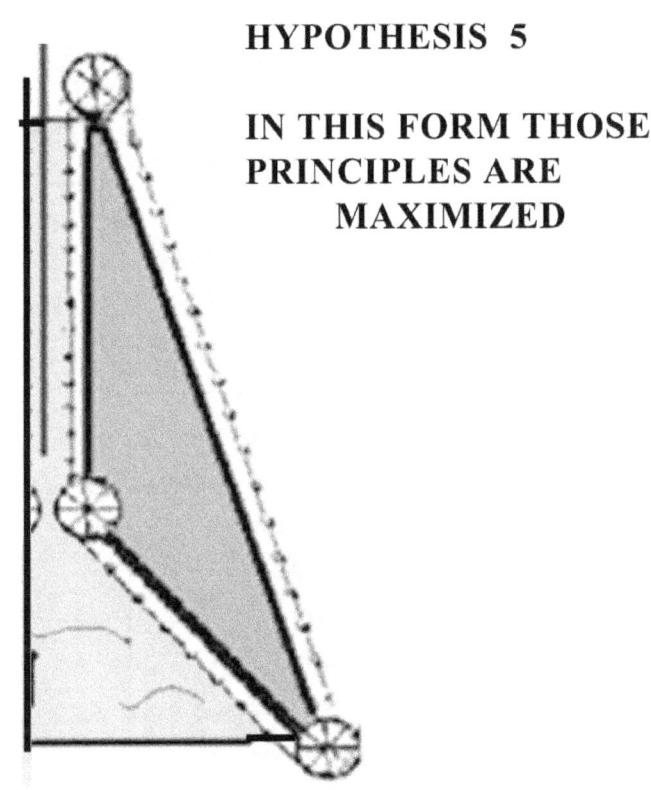

HYPOTHESIS 5

IN THIS FORM THOSE
PRINCIPLES ARE
MAXIMIZED

GRAV-BUOY DEMO 5

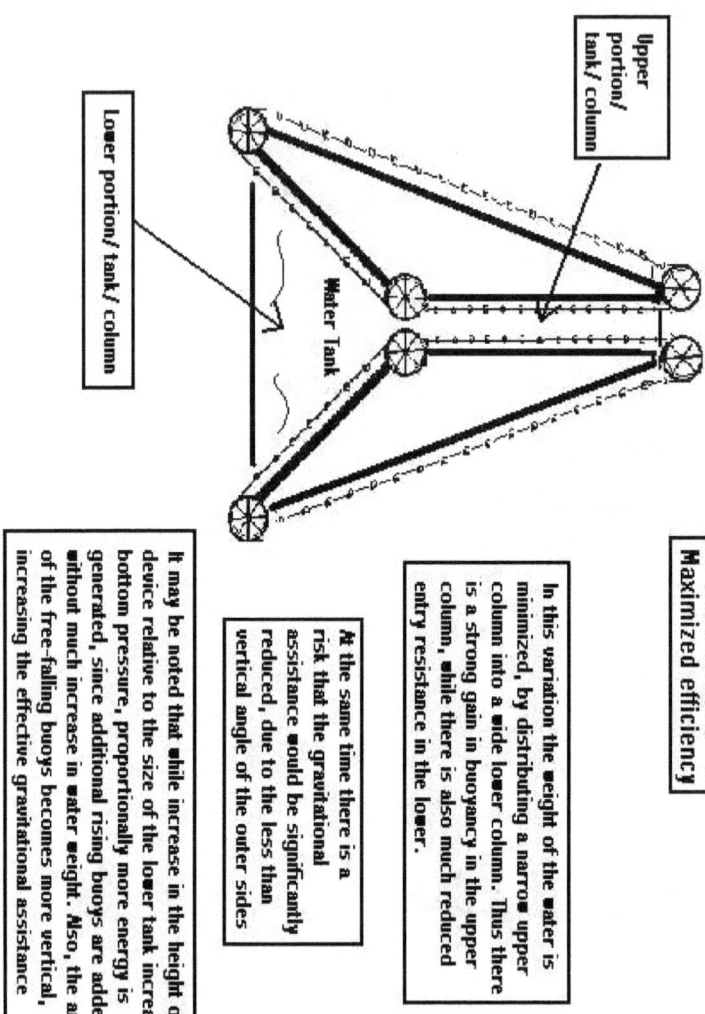

Upper portion/ tank/ column

Lower portion/ tank/ column

Water Tank

Maximized efficiency

In this variation the weight of the water is minimized, by distributing a narrow upper column into a wide lower column. Thus there is a strong gain in buoyancy in the upper column, while there is also much reduced entry resistance in the lower.

At the same time there is a risk that the gravitational assistance would be significantly reduced, due to the less than vertical angle of the outer sides

It may be noted that while increase in the height of the device relative to the size of the lower tank increases bottom pressure, proportionally more energy is generated, since additional rising buoys are added without much increase in water weight. Also, the angle of the free-falling buoys becomes more vertical, increasing the effective gravitational assistance

**FULL DESIGN FOR THE GRAV-BUOY
TYPE / ITERATION 2 (SECTION)**

FLUID LEVERAGE: THREE DESIGNS

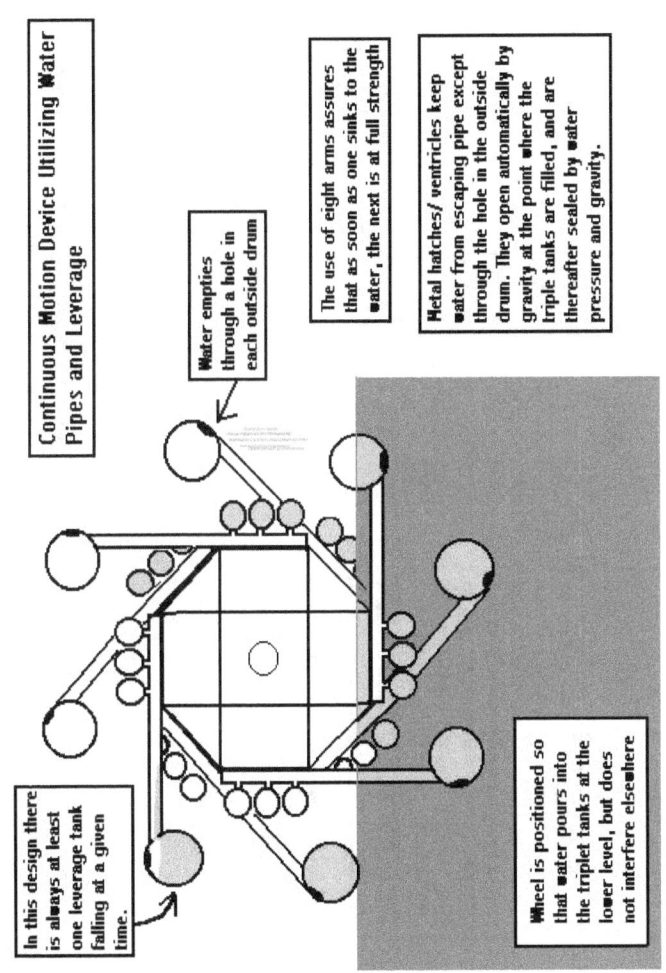

Continuous Motion Device Utilizing Water Pipes and Leverage

Water empties through a hole in each outside drum

The use of eight arms assures that as soon as one sinks to the water, the next is at full strength

Metal hatches/ ventricles keep water from escaping pipe except through the hole in the outside drum. They open automatically by gravity at the point where the triple tanks are filled, and are thereafter sealed by water pressure and gravity.

In this design there is always at least one leverage tank falling at a given time.

Wheel is positioned so that water pours into the triplet tanks at the lower level, but does not interfere elsewhere

FLUID LEVERAGE: LEVERAGE PIPES
(FIRST DESIGN)

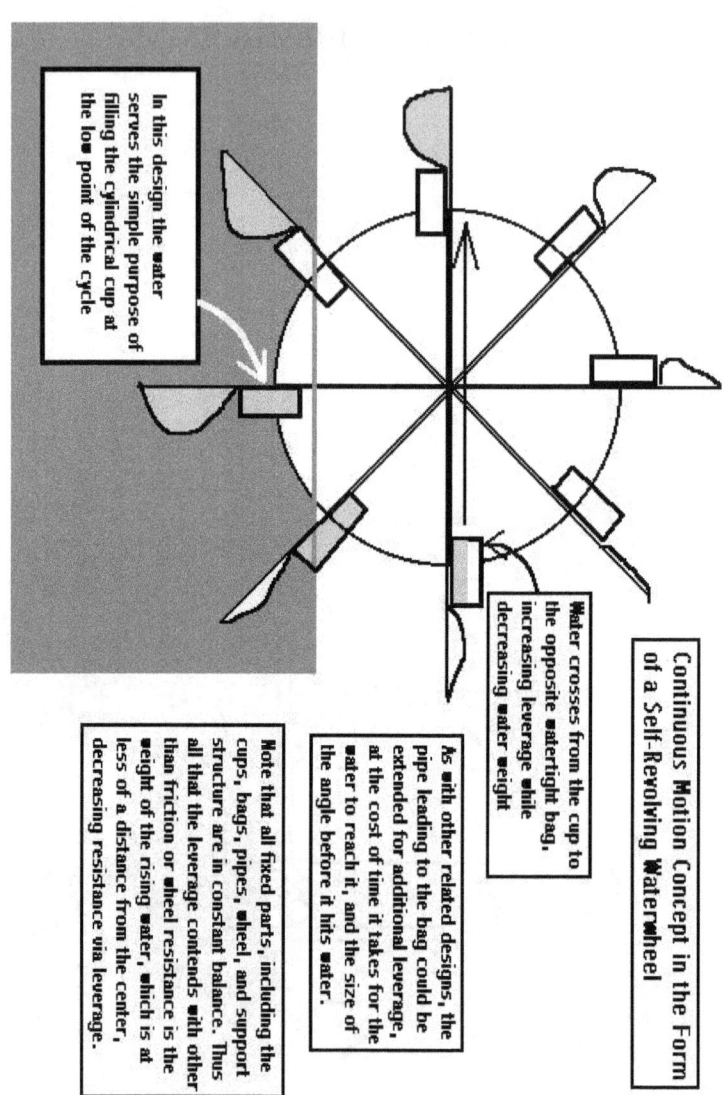

Continuous Motion Concept in the Form of a Self-Revolving Waterwheel

In this design the water serves the simple purpose of filling the cylindrical cup at the low point of the cycle

Water crosses from the cup to the opposite watertight bag, increasing leverage while decreasing water weight

As with other related designs, the pipe leading to the bag could be extended for additional leverage, at the cost of time it takes for the water to reach it, and the size of the angle before it hits water.

Note that all fixed parts, including the cups, bags, pipes, wheel, and support structure are in constant balance. Thus all that the leverage contends with other than friction or wheel resistance is the weight of the rising water, which is at less of a distance from the center, decreasing resistance via leverage.

FLUID LEVERAGE: WATERWHEEL (SECOND DESIGN, ASSUMED MEDVL)

FULLY BUOYANT WHEEL USING BUOYANT SLATS

sealed
buoyant
wheel,
partially
submerged

Here a buoyant wheel
is used, constructed in
a way where the slats
of the wheel are buoyant
as well; The wheel is
partially submerged in
water, so that there is
no resistance at the top
of the wheel; The
principle is that the
buoyancy is assymetric
at every point; if so,
it seems perpetual

buoyancy pushes up,
favoring motion only vs.
resistance

OILSKIN, BUOYANT WHEEL
(THIRD DESIGN)

CURVING RAIL DEVICE

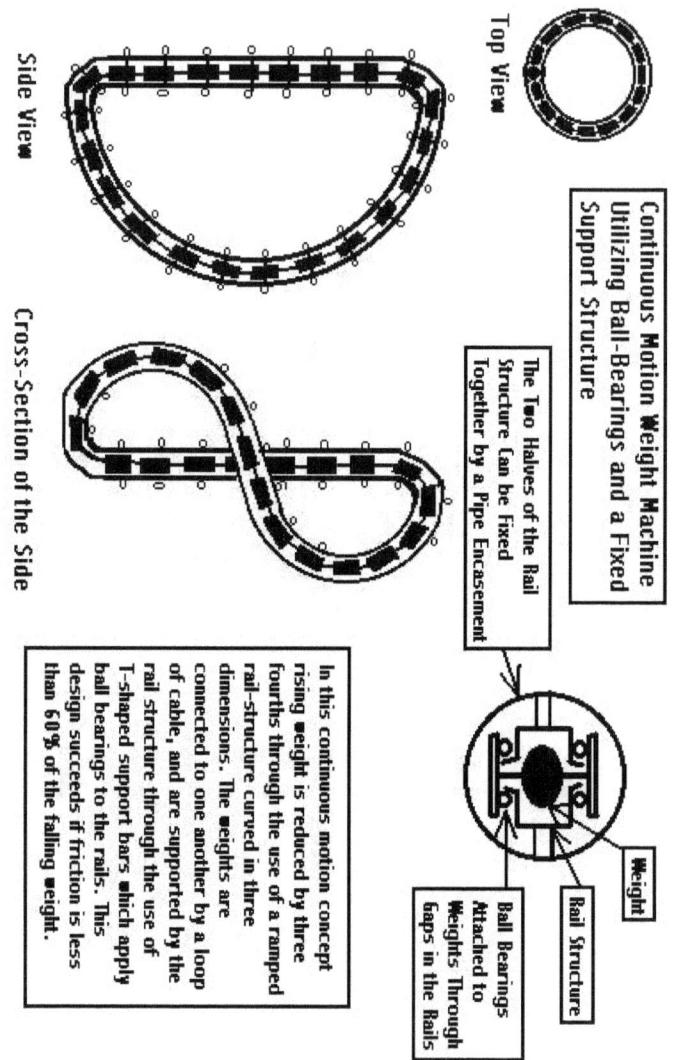

Side View

Top View

Cross-Section of the Side

Continuous Motion Weight Machine Utilizing Ball-Bearings and a Fixed Support Structure

The Two Halves of the Rail Structure Can be Fixed Together by a Pipe Encasement

Weight

Rail Structure

Ball Bearings Attached to Weights Through Gaps in the Rails

In this continuous motion concept rising weight is reduced by three fourths through the use of a ramped rail-structure curved in three dimensions. The weights are connected to one another by a loop of cable, and are supported by the rail structure through the use of T-shaped support bars which apply ball bearings to the rails. This design succeeds if friction is less than 60% of the falling weight.

CURVING RAIL DEVICE / WEIGHT MACHINE (DESIGN 1/1 FOR THIS TYPE)

MOTIVE MASS MACHINE
EXTENSIVE DESIGNS

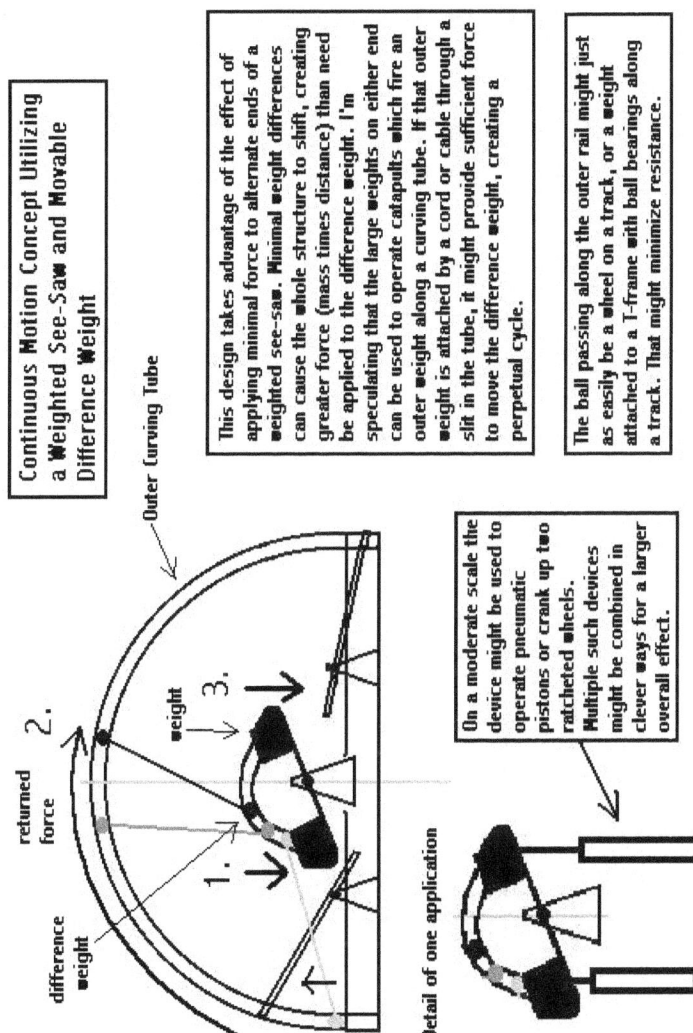

Continuous Motion Concept Utilizing a Weighted See-Saw and Movable Difference Weight

Outer Curving Tube

This design takes advantage of the effect of applying minimal force to alternate ends of a weighted see-saw. Minimal weight differences can cause the whole structure to shift, creating greater force (mass times distance) than need be applied to the difference weight. I'm speculating that the large weights on either end can be used to operate catapults which fire an outer weight along a curving tube. If that outer weight is attached by a cord or cable through a slit in the tube, it might provide sufficient force to move the difference weight, creating a perpetual cycle.

The ball passing along the outer rail might just as easily be a wheel on a track, or a weight attached to a T-frame with ball bearings along a track. That might minimize resistance.

On a moderate scale the device might be used to operate pneumatic pistons or crank up two ratcheted wheels. Multiple such devices might be combined in clever ways for a larger overall effect.

Detail of one application

returned force

difference weight

weight

2.

3.

1.

SEESAW APPARATUS
(DESIGN 1/13)

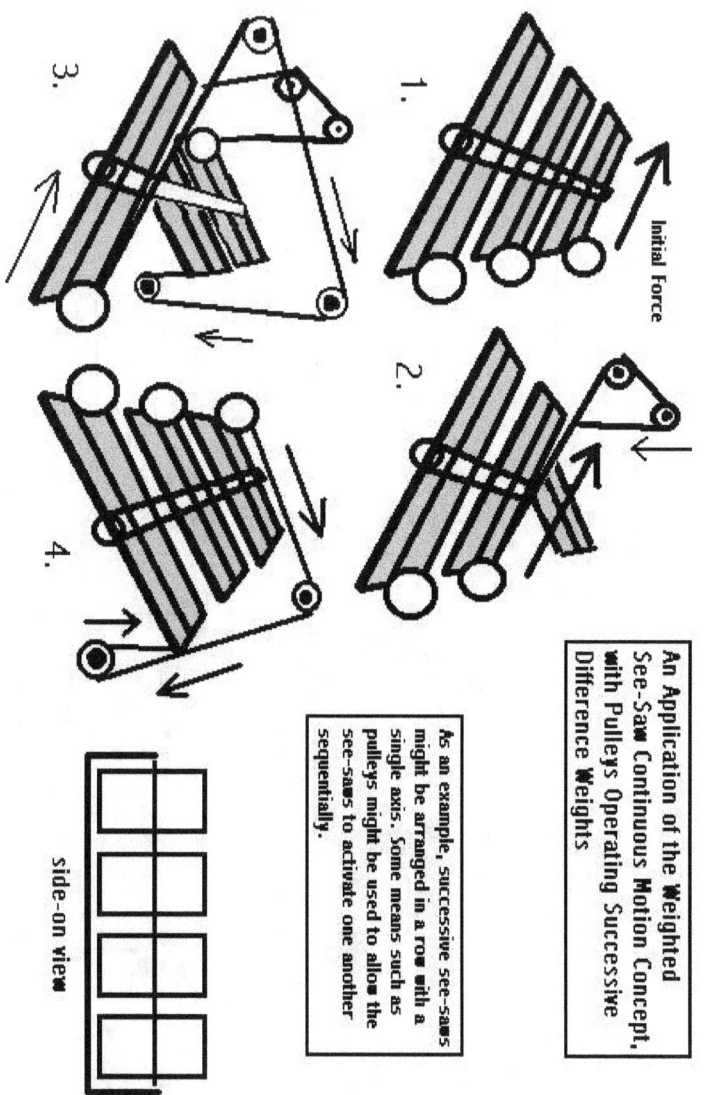

An Application of the Weighted See-Saw Continuous Motion Concept, with Pulleys Operating Successive Difference Weights

As an example, successive see-saws might be arranged in a row with a single axis. Some means such as pulleys might be used to allow the see-saws to activate one another sequentially.

side-on view

HOW TO TRIGGER A CHAIN REACTION
(BASIC RELATED CONCEPT)

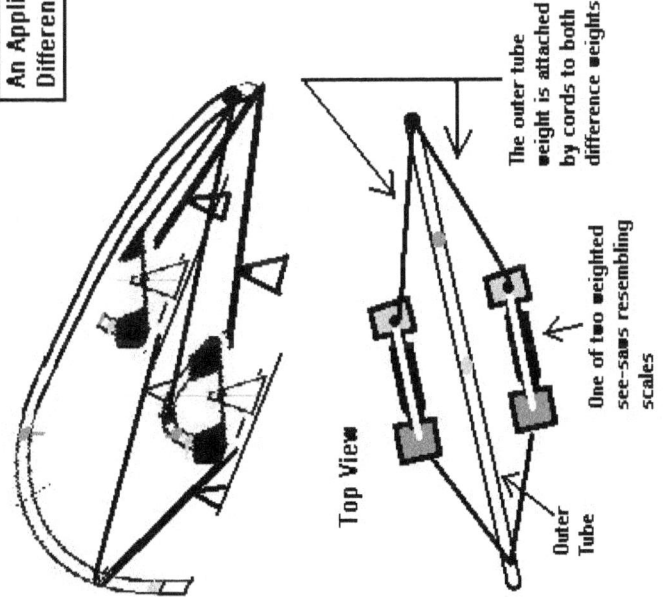

An Application of the Weighted See-Saw with Difference Weight Continuous Motion Concept

In this application two sets of weighted see-saws are used with a single outer tubed weight. The single outer weight is attached to the difference weights of both see-saws, so that slightly more than double pull is required, but provided that that pull can be generated by a single weight, two see-saws can be used more efficiently within a smaller space, leaving room for additional machinery. Doubling the number of machines active on a single catapult may permit relatively higher force applied to the outer tube weight.

The outer tube weight is attached by cords to both difference weights

One of two weighted see-saws resembling scales

Outer Tube

Top View

CATAPULTING ARCH CONCEPT
(CONCEPTUAL)

47

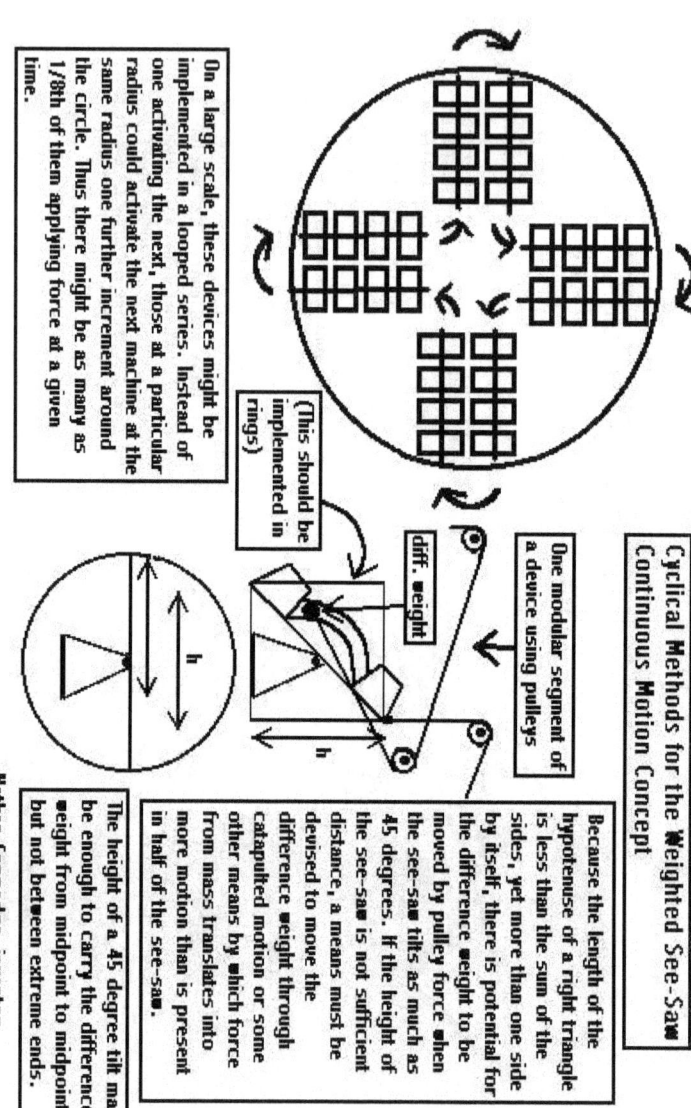

Cyclical Methods for the Weighted See-Saw Continuous Motion Concept

One modular segment of a device using pulleys

diff. weight

(This should be implemented in rings)

On a large scale, these devices might be implemented in a looped series. Instead of one activating the next, those at a particular radius could activate the next machine at the same radius one further increment around the circle. Thus there might be as many as 1/8th of them applying force at a given time.

Because the length of the hypotenuse of a right triangle is less than the sum of the sides, yet more than one side by itself, there is potential for the difference weight to be moved by pulley force when the see-saw tilts as much as 45 degrees. If the height of the see-saw is not sufficient distance, a means must be devised to move the difference weight through catapulted motion or some other means by which force from mass translates into more motion than is present in half of the see-saw.

The height of a 45 degree tilt may be enough to carry the difference weight from midpoint to midpoint, but not between extreme ends.

Nathan Coppedge, inventor

**MOTIVE-MASS CYCLICAL METHOD
SHOWING PROPORTIONAL ADVANTAGE**

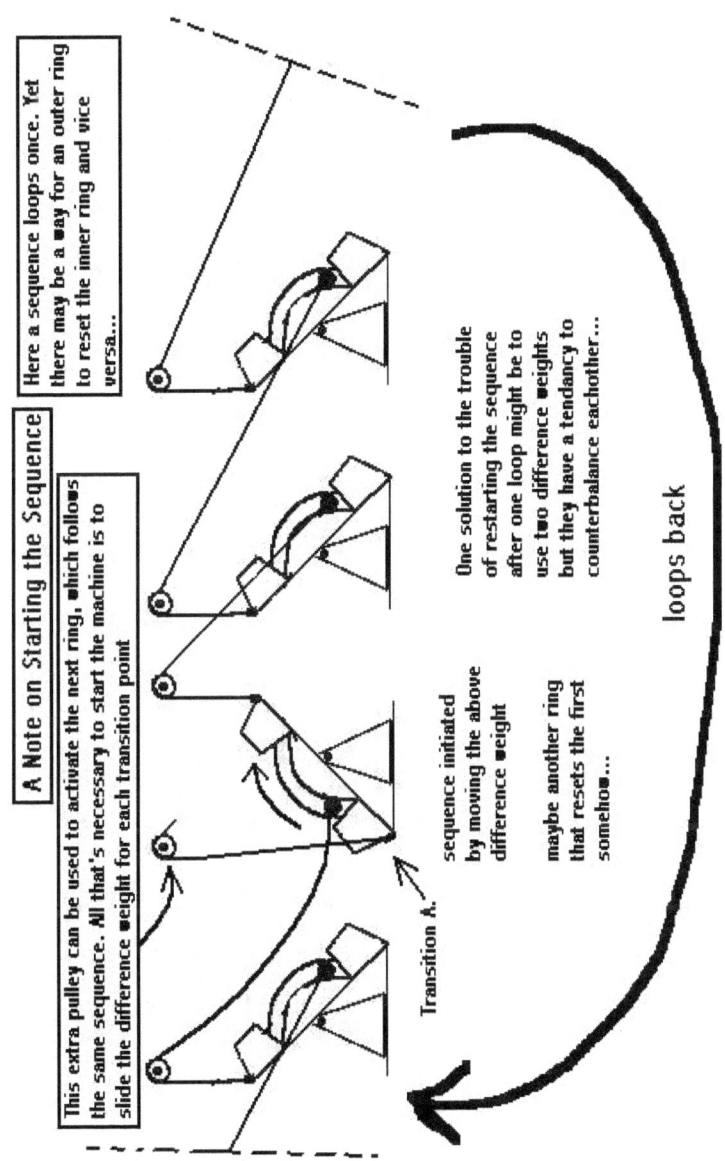

A Note on Starting the Sequence

This extra pulley can be used to activate the next ring, which follows the same sequence. All that's necessary to start the machine is to slide the difference weight for each transition point

Here a sequence loops once. Yet there may be a way for an outer ring to reset the inner ring and vice versa...

Transition A.

sequence initiated by moving the above difference weight

maybe another ring that resets the first somehow...

One solution to the trouble of restarting the sequence after one loop might be to use two difference weights but they have a tendancy to counterbalance eachother...

loops back

PULLEY CONCEPT FOR THE MOTIVE-MASS MACHINE

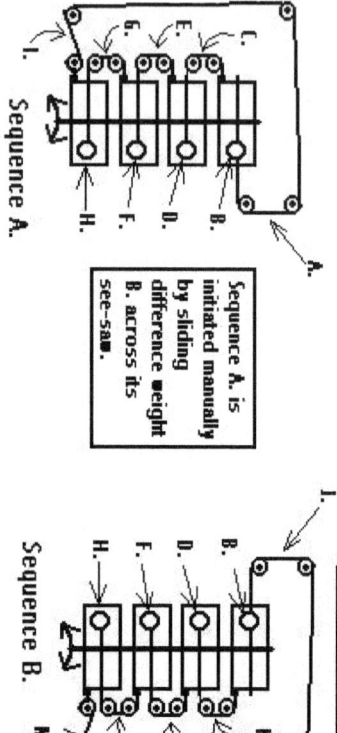

Sequence A.

The cord around pulleys A. begins loose. Diff. weight B. is moved across, tightening the cord at A. and activating the cord at pulleys C. This moves diff. weight D. across, which repeats the process by activating the cord at pulleys E, pulling F., activating G., which pulls H. H pulls the cord at I., which pulls diff. weight B. back to its initial position. However this is not by itself a self reactivating process. Note that the cord at C. is now loose since there is no pressure from B.

Sequence A. is initiated manually by sliding difference weight B. across its see-saw.

Cyclical Methods, Part II.

By combining Sequence A. and Sequence B. it seems that the process is continuous, provided that a weighted see-saw activated by a difference weight has the force to move another difference weight.

Sequence B.

Cord I. begins loose since there is no pressure from H. after Sequence A. Diff. weight B. is moved automatically by cord A. from Sequence A. loosening pulleys C. and activating pulleys K. The process repeats as diff. weight D. activates pulleys L., pulling F., activating M., pulling H. H pulls on the cord at pulleys N., activating I. Since cord A. is not receiving pressure from H., I. is free to pull on diff. weight B., reactivating Sequence A.

Nathan Coppedge, inventor

MOTIVE MASS MACHINE CYCLICAL METHOD PROCESS DIAGRAM

A rolling wheel can be pulled some distance if a free-falling equal mass is connected to it by pulley. This is because the weight of the rolling wheel is supported by the ground, whereas the falling weight transfers much of its mass into pull. If we look carefully at the diagram at left we can see that counterweight A. is supported by the track B. when pulled, but that the entire mass of A. contributes to downward pull C. through pulley B., once it has been pulled across the track.

The result is that when the weight of counterweight A. plus the smaller upper portion of the track B. is significantly greater than the larger lower portion of track B., and an initial force is applied, then one such device has the potential to activate the next in succession.

Because of the balance of masses E. and F., a significant movement of mass is possible with the application of minimal difference. The leverage force applied through the center by the counterweight allows a return on the mass of the fixed weights.

THEORY OF MOTIVE-MASS

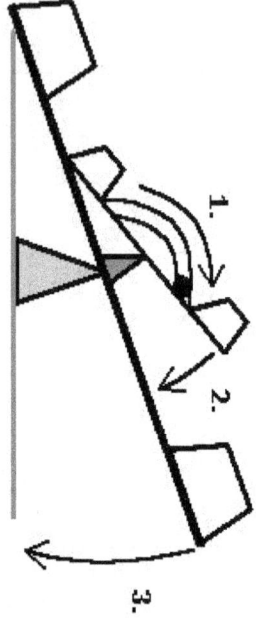

Dual Motive Mass System

In this design one motive mass see-saw is mounted on top of a larger one, thus maximizing return on the single difference weight.

The design operates in three stages per cycle. (1) The difference weight is moved. (2) The first see-saw responds to the difference weight, applying force on the larger see-saw. (3) In response, the large see-saw also tilts.

Theoretically the movement of the larger see-saw would provide more than enough force to move another difference weight, or reset its own. Note that the entire system is essentially in the same position afterwards as it was at the beginning, meaning that there is no loss of potential energy.

Application of this design in series would be similar to my most effective earlier implementation, the difference being that each unit would consist of a dual system, the difference weight pulled would be the single difference weight present here, and the larger see-saws would operate the pulleys. Note that in this design the full falling length of the larger see-saw has greater potential to move the difference weight the full distance, as it is proportionally greater than in previous designs.

DUAL-SEESAW CONCEPT

Dual Motive Mass System—Analysis of Principles

There are several things worth noting here in response to potential criticisms:

Potential criticism
"Only the difference weight contributes to the force of the larger see-saw, since the smaller see-saw is balanced"

Wrong. If the smaller see-saw can be tilted by the difference weight, the weight over time of the difference weight applied to the mass of the smaller see-saw is converted into an initial force acting from the smaller see saw to the larger one. If that force is sufficient to move the larger see-saw, we can see that the center of mass of the smaller see-saw begins to tilt. As it does so, the smaller see-saw begins to apply additional pressure on the large one by virtue of the fact that the leading weight is further along its side.

Note that the larger see-saw will begin to seek equality as soon as the left portion of the smaller see-saw moves upward. Thus the initial impact at 2. actually occurs while the larger weights are already in motion, creating a cumulative effect. That also explains how energy might be generated by the movement of the smaller see-saw: once the difference weight meets the midpoint, the see-saw is already seeking equality, so that the difference weight only adds to the effect.

DUAL-SEESAW ANALYSIS

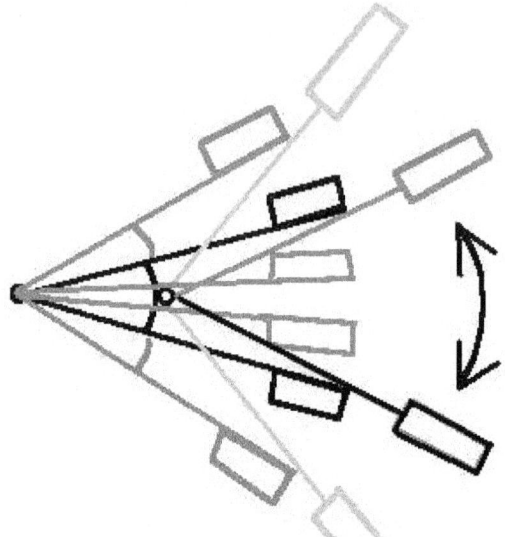

Dual Lever Motive Mass
Machine—Initial Concept

Here several different positions are depicted for the three weighted arms. The two lower arms are fixed at an acute angle from one another, and are hinged at the joining point. The upper weighted arm alternates, leaning on one of the lower weights and then the other, through use of a hinge fixed not to the acute arms, but to a separate support structure.

In principle this device works because the lower arms come into alternate distances from their axis. The majority of the upper weight is supported by the upper hinge, so that if force is provided to move it approximately 45 degrees, it then makes a return on force through the use of the acute arms, which may serve as levers.

DIFFERENCE-LEVER (DUAL LEVER) CONCEPT

MOTIVE MASS ITERATION 2

An Analysis of the Principles Operating in the Second Iteration of the Motive Mass See-Saw Concept

This is a difference weight method using a triangular track which runs upward at D. and tilts upon application of weight at C.

We can note several things about the principles operating.

1. While the length of the upward track D. is greater than the distance that a weight at C. could pull through height B., if the difference weight is pulled strictly on a horizontal, for example if the weight has an axle that runs through it with cords attached on either side of the axle, then the weight need only be pulled a lesser distance of cord along the horizontal. That distance in this case is equal to A. Since A. is equal to B. the full weight applied at C. may be transferred into moving a weight sufficient distance to perpetuate the process.

Nathan L. Coppedge

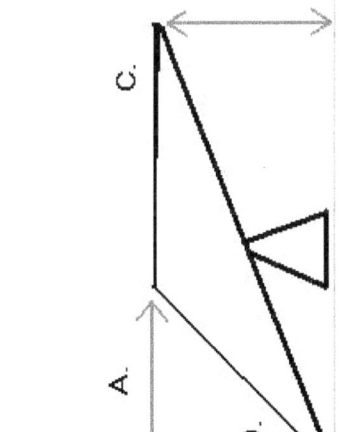

2. In this particular geometrical configuration, the length of the track at C. is actually sloped downward even when it is at its greatest height. This allows the difference weight to roll the remaining distance of the see-saw after it has reached the midpoint of the track.

MOTIVE MASS IT2 THEORY

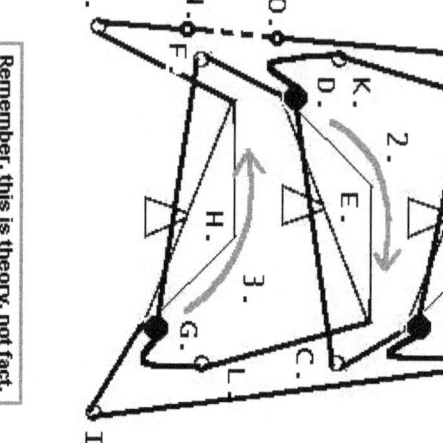

Remember, this is theory, not fact.
I haven't built this device.

An Application of the Second Iteration of the Motive Mass See-Saw Concept in a Vertical Series, Consisting of Two Cycles

An over-unity method consisting of three specially built see-saws with rolling difference weights, connected by pulleys.

Cycle I

The first cycle begins by manually moving difference weight A. towards pulley P. The force of the weight is transferred through see-saw B. to a cord attached on the right side of see-saw B. through pulley C., pulling difference weight D. As a result, force is transferred through see-saw E., across pulley F., pulling G. across H., transferring force through I. and J., pulling A. back across.

Cycle II

Force is then transferred through pulley K., moving D. back across, transferring force through L., moving G. back across. See-Saw H. then moves A. across for a third time, through pulley P. Since the fastening between O. and N. has been attached during the first cycle, difference weight A. is now sealed into the cycle as much as D. and G.

Nathan L. Coppedge

FINAL DESIGN FOR MOTIVE MASS (ITERATION 2)

CHAIN-REACTION TOWER CONCEPT

Chain Reaction Tower
Perpetual Motion Concept

Difference lever (A.) is toggled, excentuated by stouter difference wedge (B.), operating a similar longer difference wedge (C) via a bracket; Process continues in the blockier lower structure by a domino principle; Cycle resets through operation at bracket (D.) and cross-beams at (E.); The three sets are designed to move in alternating directions, creating a chain reaction::

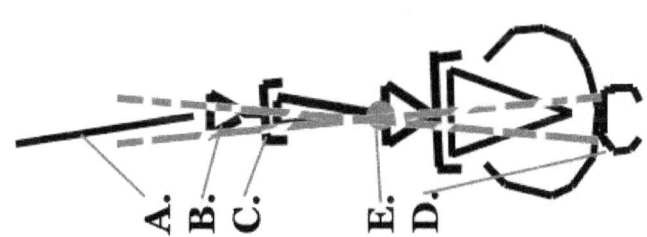

Nathan L. Coppedge

MOTIVE-MASS CHAIN-REACTION TOWER

REPEATING LEVERAGE APPARATUS

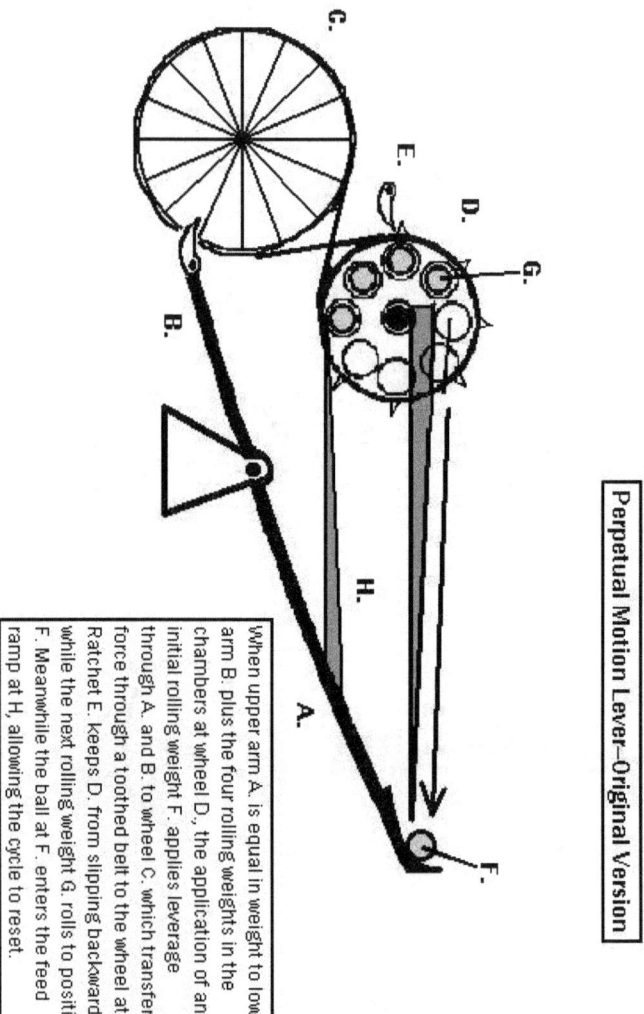

Perpetual Motion Lever–Original Version

When upper arm A. is equal in weight to lower arm B. plus the four rolling weights in the chambers at wheel D., the application of an initial rolling weight F. applies leverage through A. and B. to wheel C. which transfers force through a toothed belt to the wheel at D. Ratchet E. keeps D. from slipping backward while the next rolling weight G. rolls to position F. Meanwhile the ball at F. enters the feed ramp at H, allowing the cycle to reset.

DESIGN / ITERATION 1

Repeating Leverage Concept Using a Counter-Weight and Slopes Leading to a Free-Fall at a Point of Greater Leverage

Below: A Side-On View of how the central leverage structure in fact passes between two triangles, thus allowing the rolling cylinder to be supported at every point in its movement. As depicted the counter-weight is on the other side. Protruding ends of the rolling cylinder are shown in order to give an idea of its location.

Below: The counter-weight at A. is meant to cause the cylindrical rolling weight B. to gradually rise from point C. to point D. Since point D. is a point of minimum leverage for rolling weight B., it would be forced upward on the slope from D. to E., where its weight is partially supported, reducing the leverage applied to the counter-weight. At point E. the cylinder B. rolls of the track, causing its whole weight to act at greater leverage until it returns to point C. Because there is a slope upward to D., the direction of least resistance is to follow the cycle once more.

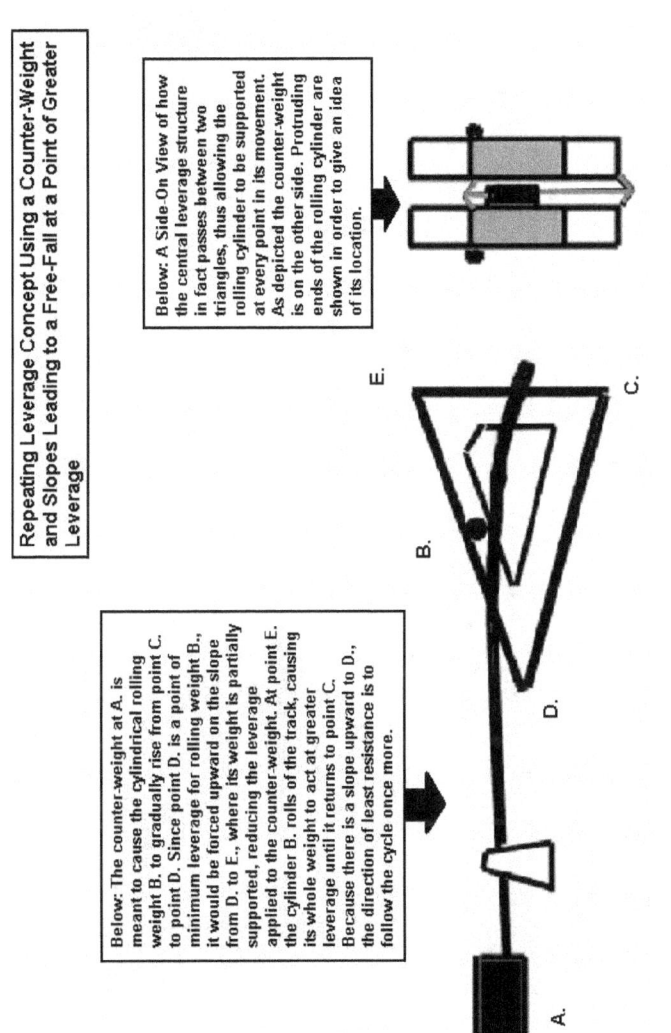

Nathan L. Coppedge

REPEATING LEVERAGE TYPE 2

Repeating Leverage Concept Using a Counter-Weight and Slopes Leading to a Free-Fall at a Point of Greater Leverage—Variation 1

counterweight

A.

lever, shown in three positions

level

D.

B.

E. rolling weight

C.

Side View

lever — center support — track

A modified design using a longer lever, with the virtue that the angle remains sloped towards the vertical drop for the entire length from D. to E., while the angle also remains sloped towards the point of least leverage from B., all the way from C. to D.

The only trouble may be in finding a weight value for counterweight A. by which the vertical drop from E. to C. involves sufficient leverage from B. to move the counterweight upward, while the short rise between C-D and D-E involves a weight value from B. that is sufficiently less than the weight value of counterweight A. so that the leverage at that point directs B. upwards.

This might be accomplished by modifying the length of the counterbalance shaft and weight value, which do not depend on the angle of the lever, and may be considered variable.

Note that the angle of D. can be made more acute, as long as it lies between the uppermost and lowermost points of the lever, indicated by the nearly horizontal black and light gray lines extending from the hinged support for the lever. This has the effect of improving the ratio between the leverage applied at the vertical drop and the minimum of leverage necessary to lift the weight back at the point (D.) of lowest leverage from the rolling weight, and may be made to approximate 2:1 or even 3:1, which should be sufficient.

Nathan L. Coppedge

TYPE 2, VARIATION 1

Repeat Lever Type 3
Eucaleh Terrapin, Inventor

Lever passes
in slot through
guides for
descending
ball weight

A.

B.

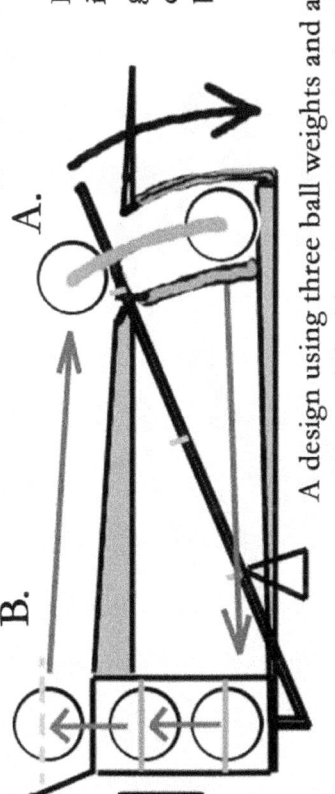

A design using three ball weights and a double chambered structure lifted by the short end of a lever. According to the proportions as pictured the movement of a single weight at A. would be sufficient to lift both chambered weights, the topmost weight being incremented onto a gentle ramp at B. towards point A., the bottom-most weight being incremented into the upper chamber through a side ramp (not pictured), and the entire operation facilitated by a counterweight by pulley, equal to slightly less than one chambered weight (one chambered weight being constant)

REPEAT LEVER TYPE/ ITERATION 3

Repeat Lever Type 4   Eucaleh Terrapin, Inventor

A method in which two ball weights follow a variable path, the upward portion extending in a crescent followed closely by a mobile guide bar shaped to support one ball weight along the upward grade and attached to the longer end of a lever,

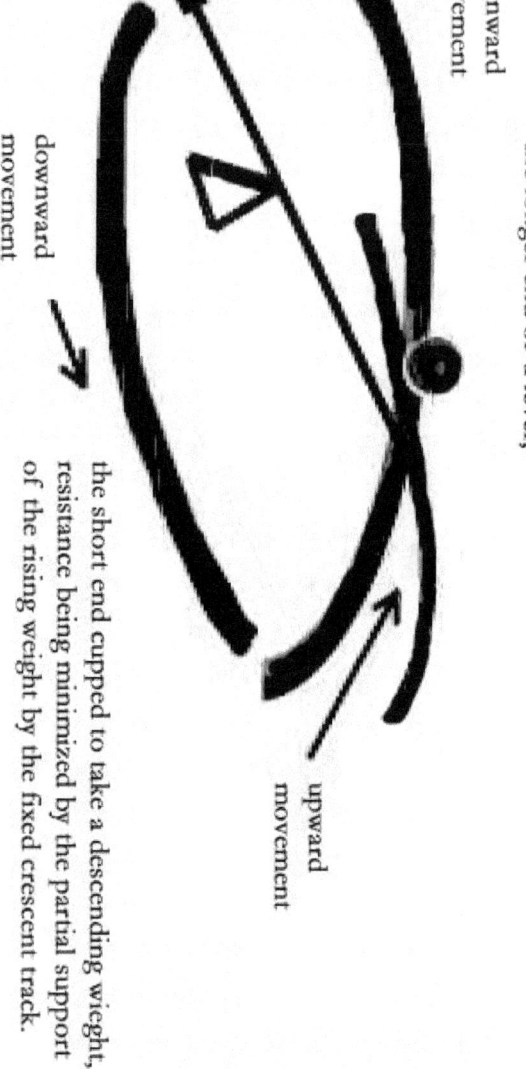

downward movement

downward movement

upward movement

the short end cupped to take a descending wieght, resistance being minimized by the partial support of the rising weight by the fixed crescent track.

REPEATING LEVERAGE TYPE 4

ARCHEMECHANICS: Trough Lever Device

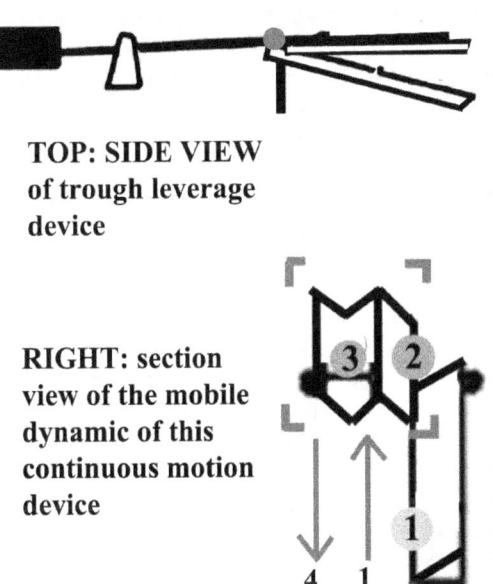

TOP: SIDE VIEW of trough leverage device

RIGHT: section view of the mobile dynamic of this continuous motion device

Double-trough device in which a partial trough assists upward leverage via a fixed half-track:

The dynamic is meant to promote energy generation by movement of a fulcrum, since greater weight bears on the mobile track when the ball weight is not supported by the fixed track segment

(c) 2011 Nathan Coppedge

REPEAT LEVER TYPE / ITERATION 5:

BELIEVED TO BE FUNCTIONAL IN SOME FORM: MAKES USE OF GREATER WEIGHT BEARING ON MOBILE THAN FIXED APPARATUS, USES RAMPS TO DELIVER WEIGHT IN AND OUT OF APPARATUS

FRONT VIEW

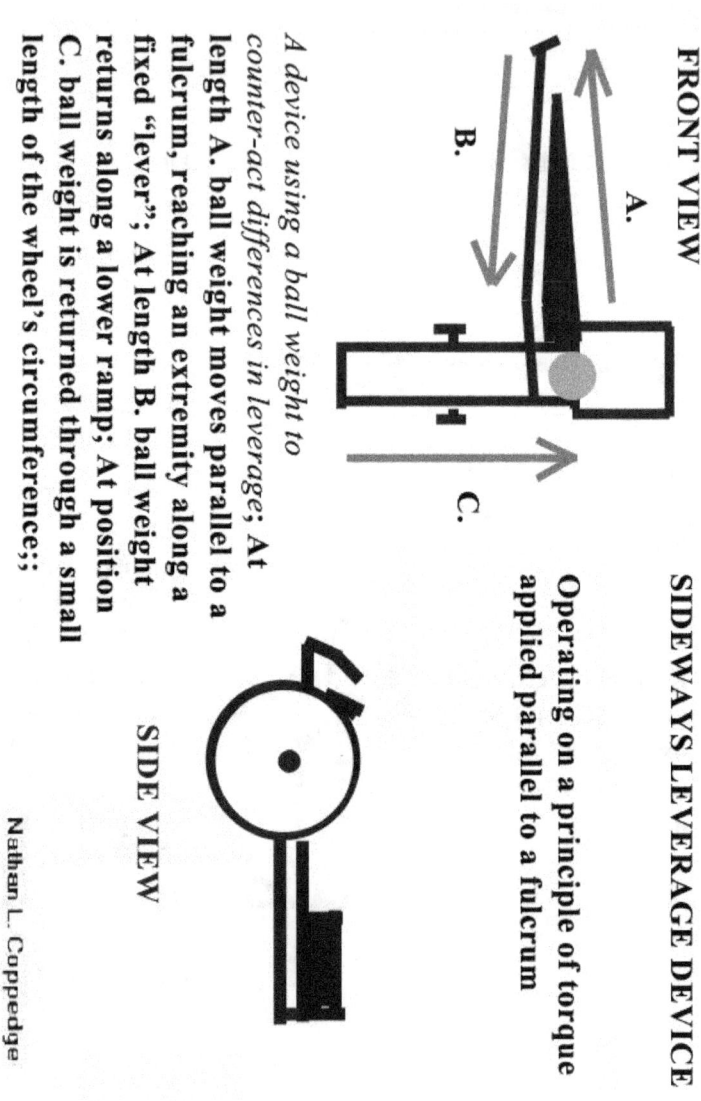

A.

B.

C.

SIDEWAYS LEVERAGE DEVICE

Operating on a principle of torque applied parallel to a fulcrum

SIDE VIEW

A device using a ball weight to counter-act differences in leverage; At length A. ball weight moves parallel to a fulcrum, reaching an extremity along a fixed "lever"; At length B. ball weight returns along a lower ramp; At position C. ball weight is returned through a small length of the wheel's circumference;;

Nathan L. Coppedge

**SIDEWAYS LEVERAGE DEVICE:
A FAILED CONCEPT (HAS BEEN
TESTED)**

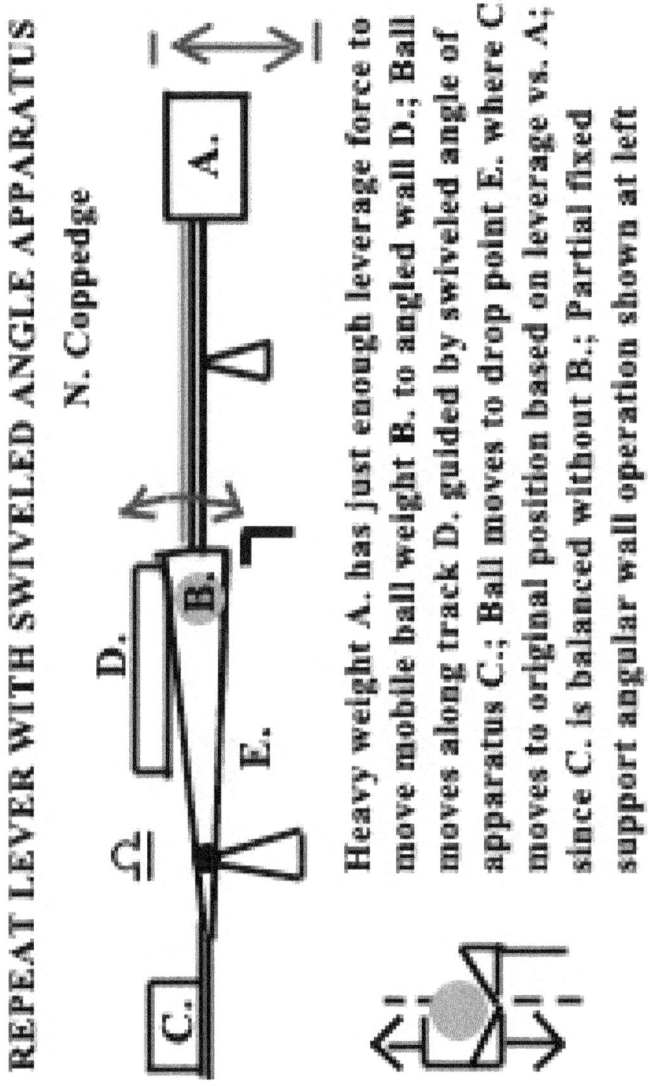

REPEAT LEVER WITH SWIVELED ANGLE APPARATUS

N. Coppedge

Heavy weight A. has just enough leverage force to move mobile ball weight B. to angled wall D.; Ball moves along track D. guided by swiveled angle of apparatus C.; Ball moves to drop point E. where C. moves to original position based on leverage vs. A; since C. is balanced without B.; Partial fixed support angular wall operation shown at left

COMPLICATED VARIATION OF THE REPEAT LEVERAGE APPARATUS

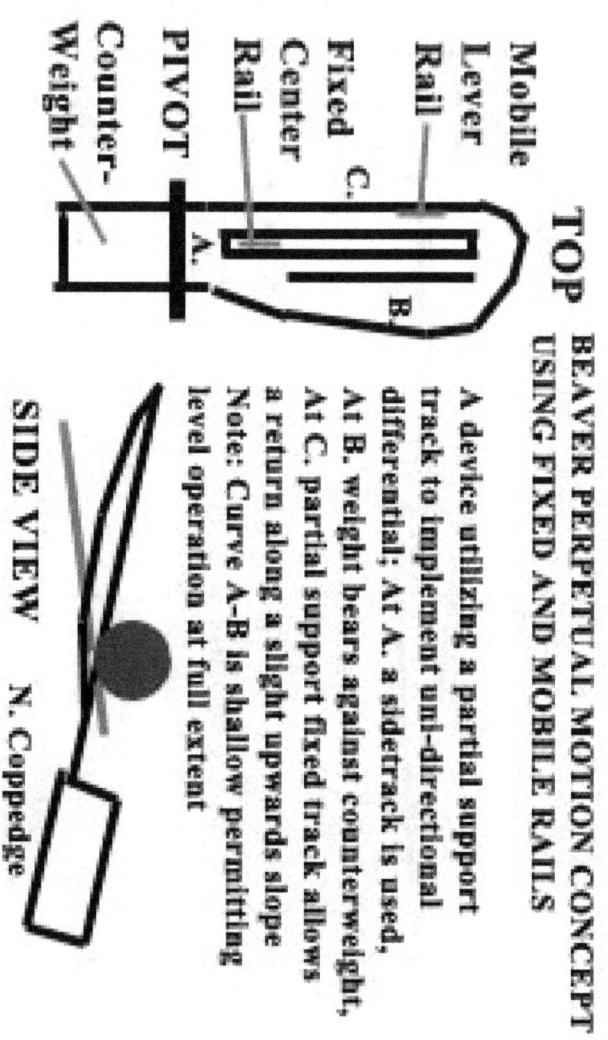

TOP

BEAVER PERPETUAL MOTION CONCEPT
USING FIXED AND MOBILE RAILS

Mobile Lever Rail

Fixed Center Rail

PIVOT

Counter-Weight

A device utilizing a partial support
track to implement uni-directional
differential; At A. a sidetrack is used;
At B. weight bears against counterweight,
At C. partial support fixed track allows
a return along a slight upwards slope
Note: Curve A–B is shallow permitting
level operation at full extent

SIDE VIEW   N. Coppedge

**BEAVER TYPE / FIXED RAIL SUPPORT
VARIATION OF THE TROUGH LEVER**

TROUGH DEVICE USING SPIRAL-CURVED SLOPE

A device modifying an earlier principle for gain on differential support using a counterweight, in which as earlier a single ball weight is intended to operate, first upon a full trough, lacking support and therefore taking full weight of the mobile ball weight, and secondly upon a mobile half-trough partially supported by a fixed secondary half-trough

Nathan L. Coppedge

SIDE VIEW

counterweight

During the cycle, the ball weight moves from position A1 and A2, where the trough takes the weight, to positions B1 and B2 where the half-trough provides support

1.5 troughs, curved and mobile

A2

A1

B1

B2

TOP VIEW

fixed half-trough

arrows show cycle

TROUGH LEVER VARIATION USING A CURVED LEVER

N. COPPEDGE

VARIATION ON THE
REPEAT LEVER
CONCEPT

ALTERNATION
BETWEEN
PARTIAL
SUPPORT AND
DOWNWARDS
SLOPE

**SECANT TYPE REPEATING LEVER
APPARATUS**

DIFFERENCE PENDULUM DEVICE USING SPIRAL SLOPE AND STEEP OFFSET DROP POINT AND RETURN DRAG

At drop point (A) mobile ball weight has advantage against counterweight (B); However, according to design when ball weight reaches point (C) it is supported by a spiral track, decreasing effective weight, and causing counterweight (B) to act by dragging the weight upwards; By outwards angularity of the track and upwards pull, the ball weight is thrown in a spiral, until slope allows its full weight to apply at (C)::

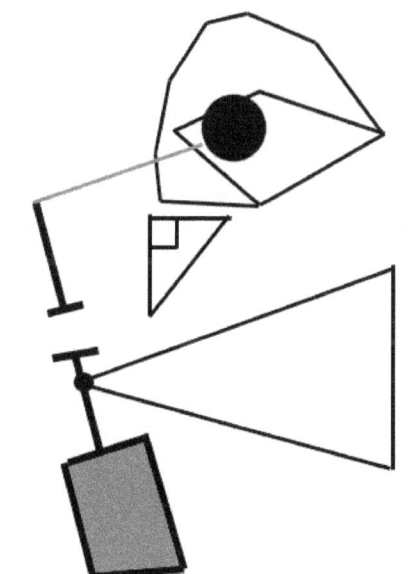

Nathan Coppedge

REPEATING LEVERAGE VARIATION USING A "DIFFERENCE SPIRAL"

REPEAT LEVER EXPERIMENT
It is clear from this experiment that motion is
not the sole factor of continuous momentum;
For rolling ball at point (A) CAN return to point
(C) at greater altitude, if it is allowed that the
activated lever returns to a greater altitude;
Again, the effectiveness of continuous
momentum depends on specifics

(Approx. 2/3rds lever)

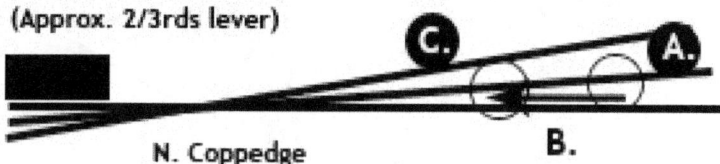

N. Coppedge   B.

REPEATING LEVERAGE EXPERIMENT

REPEATING LEVERAGE PERPETUAL MOTION MACHINE USING A HINGED STRUCTURE AND CLEVER USE OF RAMPS

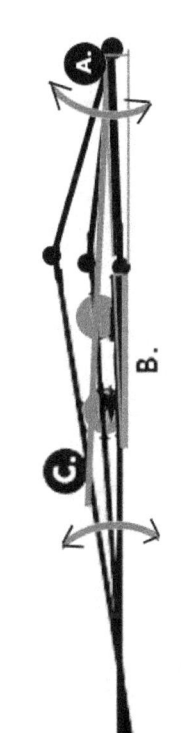

Nathan L. Coppedge

A counter-weight with approximately 3X mass at 1/2 hinge distance allows this machine to operate, when the structure is lightweight. In other cases both the counterweight and ball weight must be heavy to accomodate a heavy structure

Repeating leverage device making use of special ramp applications and a counterweight versus leverage application.

At point (A) rolling ball applies high leverage, operating the hinged lever structure and moving along section (B) which is sloped downwards. At the point beyond the hinge, the counterweight begins to exert greater force, and the ball weight is thrown upwards, at (C). A side ramp at (C) permits the ball weight to return to point (A) where the entire structure of the apparatus has returned to the initial position.

REPEATING LEVERAGE VARIATION USING A HINGE TO INCREASE SUPPORT-TO-DISTANCE

MODULAR TROUGH LEVERAGE

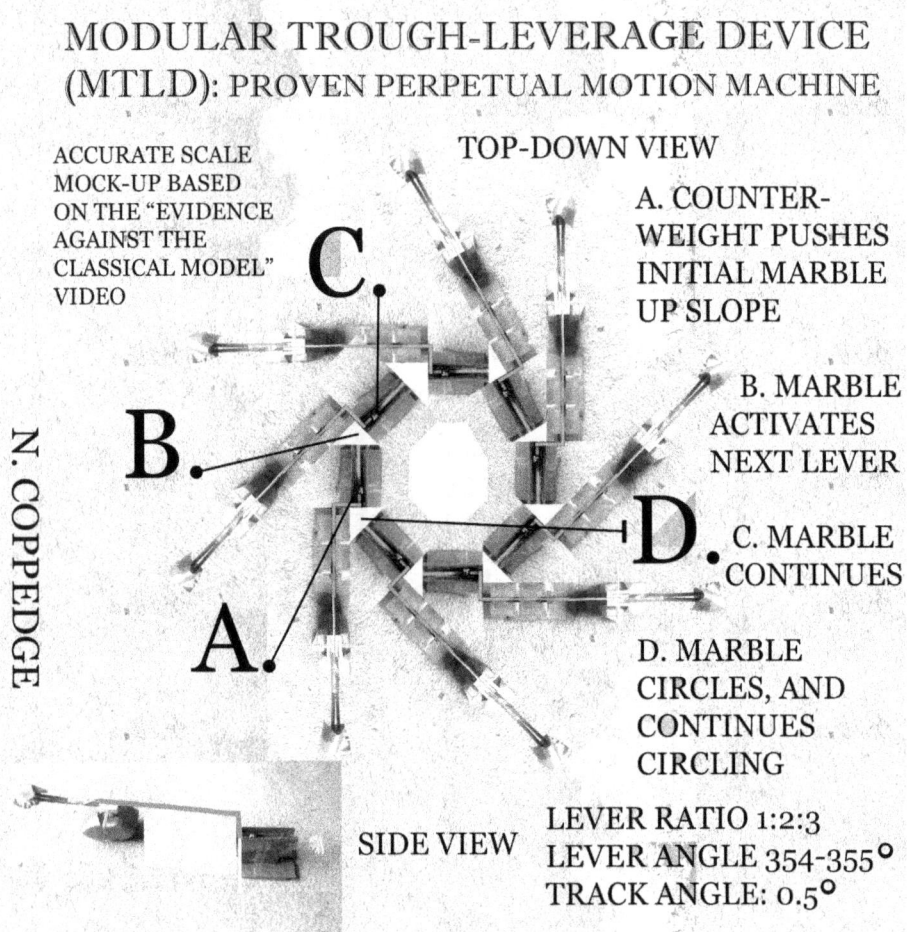

MODULAR TROUGH-LEVERAGE DEVICE
(MTLD): PROVEN PERPETUAL MOTION MACHINE

ACCURATE SCALE
MOCK-UP BASED
ON THE "EVIDENCE
AGAINST THE
CLASSICAL MODEL"
VIDEO

TOP-DOWN VIEW

A. COUNTER-
WEIGHT PUSHES
INITIAL MARBLE
UP SLOPE

B. MARBLE
ACTIVATES
NEXT LEVER

N. COPPEDGE

C. MARBLE
CONTINUES

D. MARBLE
CIRCLES, AND
CONTINUES
CIRCLING

SIDE VIEW

LEVER RATIO 1:2:3
LEVER ANGLE 354-355°
TRACK ANGLE: 0.5°

PROVEN PERPETUAL MOTION CONCEPT

TILT MOTOR

Tilt-Motor Perpetual Motion Concept

Original concept for a rotary device in which a weighted cone rolls around a swivel, activating successive pressure plates or "keys" operating levers. The levers in turn apply upwards pressure at a 90 degree angle on the track behind the cone. Since the track swivels downwards towards the portion weighted with the rolling cone, the upwards pressure is designed to create a continuous slope which follows the cone as it activates successive pressure plates.

Because the pressure plates are located outside the perimeter of the track, the cone's weight on the "wickets" on the active end of the levers only causes the pressure plate keys to be raised, rather than inhibiting movement by causing conflictive movement of the track. Metal "steps" are attached to the pressure plate keys in order to assure that the cone is activating one pressure plate at a given moment, which is meant to be sufficient to allow continuous motion.

Nathan L. Coppedge

Top-Down View

Side View

THIRD-BEST OPTION FOR
PERPETUAL MOTION

THE ESCHER MACHINE

C. Master Angle 2: marble rolls upwards again, using a differently-directed master angle

D.
Ramp 2: Using altitude from Master Angle 2, marble returns to Master Angle 1

B.
Ramp 1: A down-wards motion is possible due to the gain in height

A. Master Angle 1: marble rolls upwards using a horizontal slope

NATHAN COPPEDGE

BEST OPTION FOR PERPETUAL MOTION

THE ESCHER MACHINE

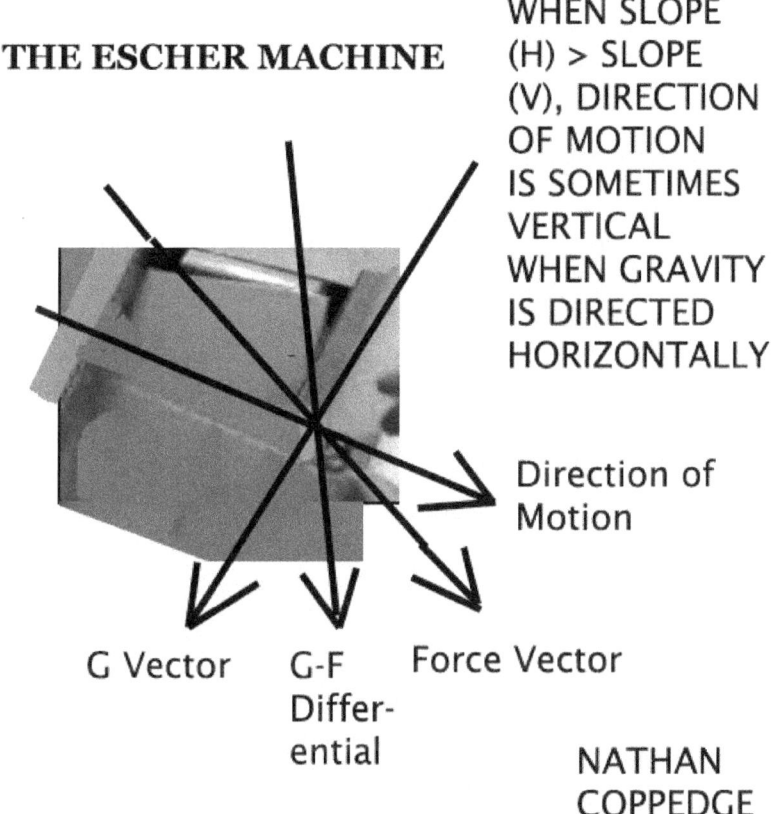

WHEN SLOPE (H) > SLOPE (V), DIRECTION OF MOTION IS SOMETIMES VERTICAL WHEN GRAVITY IS DIRECTED HORIZONTALLY

Direction of Motion

G Vector

G-F Differ-ential

Force Vector

NATHAN COPPEDGE

ESCHER MACHINE PHYSICS

Nathan Coppedge

COQUETTE

IDEAL COQUETTE

A CONCEPT I DREAMED UP IN 2007

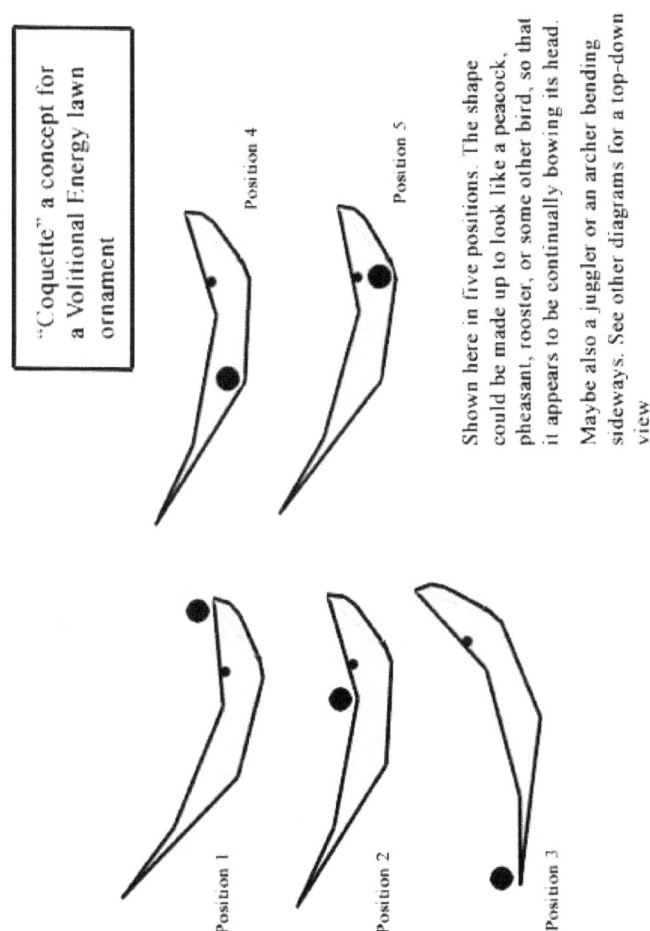

"Coquette" a concept for a Volitional Energy lawn ornament

Position 4

Position 5

Shown here in five positions. The shape could be made up to look like a peacock, pheasant, rooster, or some other bird, so that it appears to be continually bowing its head.

Maybe also a juggler or an archer bending sideways. See other diagrams for a top-down view

Position 1

Position 2

Position 3

**OPERATION OF THE COQUETTE
(THEORETICAL)**

COQUETTE: A Volitional Energy Lawn Ornament Concept
Eucaleh Terrapin, Inventor

In this device, which may function as a lawn ornament, or if suitable in that role, also as a mechanical device, a structure of track or tubing is built over a fulcrum, like a see-saw divided into three parts that meet at each end.

One end is considerably shorter than the other, that is so that Point A. is much closer to the fulcrum than Point B. A single ball-bearing type weight is used, so that when it is placed at Point A the natural slope leads downwards some ways towards Point B. When the ball bearing reaches a point between A. and B where its weight would account by leverage for the shorter end of the pivoting structure, the angle is such that if it did not the slope would be prohibitively steep, but instead the weight of the ball-bearing extends the slope until it reaches point B.

At Point B the ball bearing may take one of two routes, which are identical. It may follow 1 or 4 downwards some length until Point B. begins to rise again. Then at 2 or 5 the slope is such that it would be prohibitive if Point B were at its lowest. Finally at Seg. 3 or 6 the slope is down into Pt. A

A Top-Down View

Point B.

Seg. 1

Seg. 2

Seg. 3

Point A.

Seg. 4

Seg. 5

Seg. 6

Fulcrum

Note: because the slope averages an upwards grade between Point A and Point B, if the device did not pivot, by the time the ball bearing returns to Point A. it may travel downwards.

PRIMARY OPERATION OF THE COQUETTE

Coquette Type 2

Tilting device relying upon a differentiation in leveraged weight application, vis. at Point A the heavy weight is resisted since, although it has greater bearing, it applies less leverage than the counterweight, however at point B the light weight is resisted via strong counterposed leverage combined with a heavy weight at less leverage distance than the rolling ball. By following a circular or ovular track the ball weight thus finds points within which a forward impetus is reinforced by the tendency of the device to tilt where the ball bears down, while returning with force to an anteceding position; the principle of light leverage (counterpointed by leverage application) counterpoints heavy weight application at a point of minimal leverage.

|| Eucaleh Terrapin ||

A.

light counterweight
w/ leverage

heavy weight with
low leverage / torque

circular /
ovular track

b.

this design is based on bisected metal supports for atmospheric lights

TECHNIQUES

LEFT: key slope
points corresponding
to major lever applications

COQUETTE TYPE 2 (NOTE DISK)

PERPETUAL MAGNET CONCEPTS

N. Coppedge

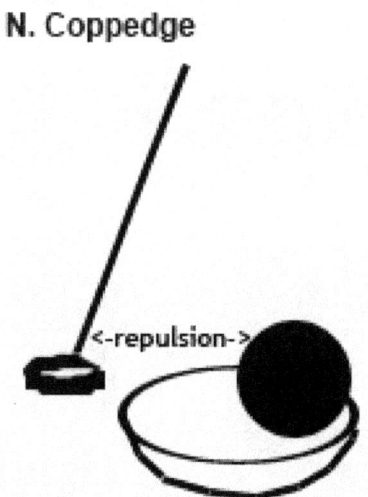

MAGNET DESIGN 1

A circular magnet is attached at its midpoint to a tether, above a spherical bowl in which a larger metal ball is made to roll; When the magnet approaches the ball, the ball rolls, applying its weight to rotation

MAGNET MACHINE TYPE 1: HORIZONTAL APPLICATION OF WEIGHT AGAINST AN UPWARDS CURVE IS DESIGNED TO CREATE A PERPETUALLY-ROTATING PENDULUM. NOTE THE HORIZONTAL CURVE OF THE BALL IN THE MIDDLE

[SEE ALSO GRAVITY MOTORS, LATER]

N. Coppedge

PERPETUAL MOTION MAGNET 1

Modification

If a single magnet is both attractive and repulsive, two magnets can be mounted together oppositely in a cylindrical form, with the mid-point being aligned with the mid-point of the larger ball weight, creating the same effect of repulsion

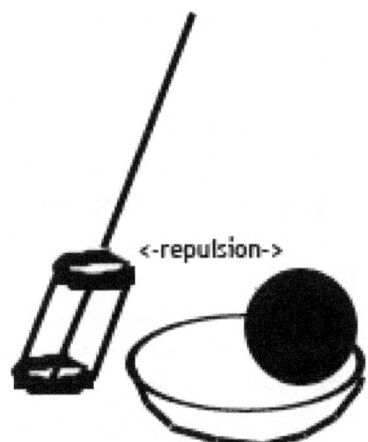

<-repulsion->

MAGNET MACHINE TYPE 1, VARIATION 1: COMPRESSING MAGNETS TOGETHER MAY GENERATE MORE OR EXCESS FORCE, BUT RUNS THE RISK OF DEGRADING THE MAGNETS

BEZEL WEIGHT DEVICE

TRACKED APPARATUS USING A BEZEL-WEIGHT AND
ROLLING BALL

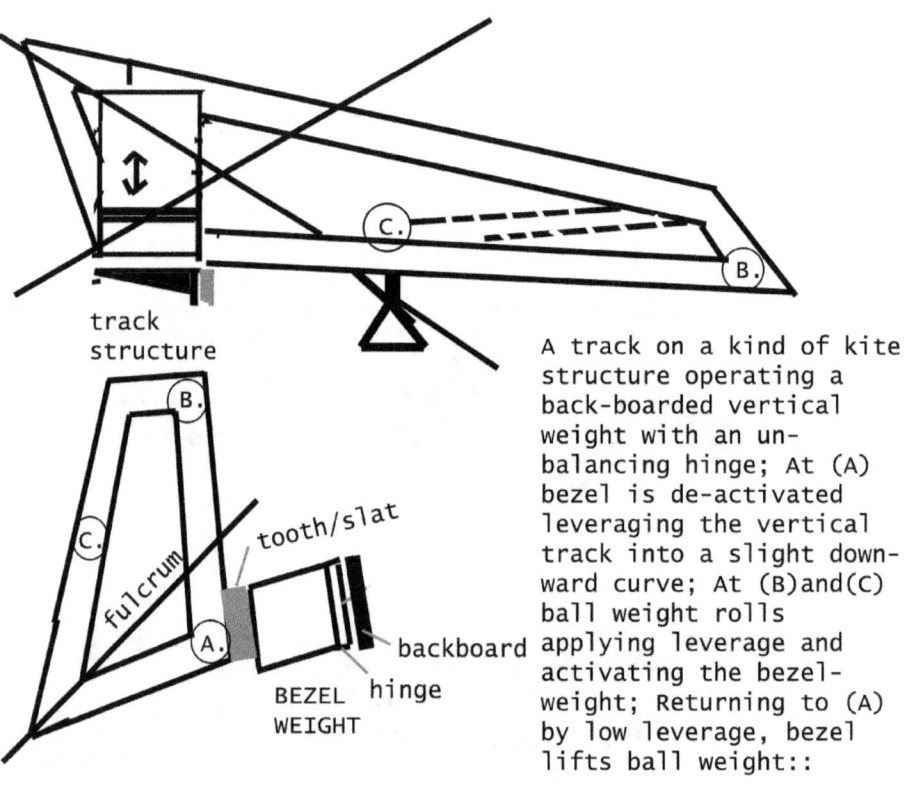

track
structure

A track on a kind of kite
structure operating a
back-boarded vertical
weight with an un-
balancing hinge; At (A)
bezel is de-activated
leveraging the vertical
track into a slight down-
ward curve; At (B)and(C)
ball weight rolls
applying leverage and
activating the bezel-
weight; Returning to (A)
by low leverage, bezel
lifts ball weight::

BEZEL WEIGHT TYPE 1/1

BEZEL DEVICES MAKE USE OF A WEIGHT
AGAINST A BRACKET TO CREATE SPRING-
WORK INDEFINITELY

GRAVITY MOTOR

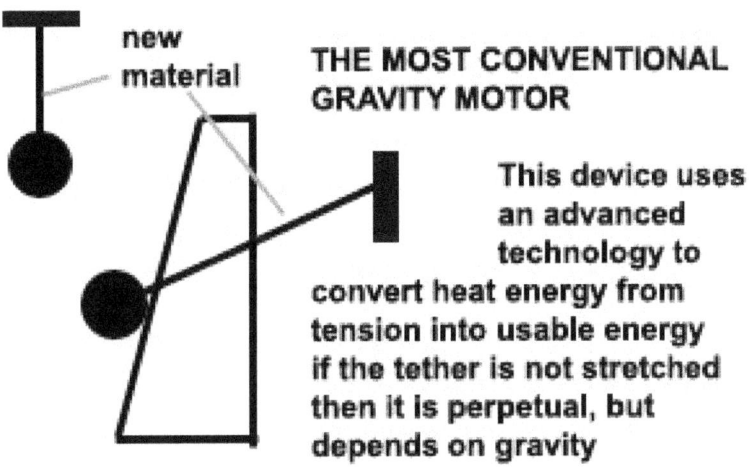

new material

THE MOST CONVENTIONAL GRAVITY MOTOR

This device uses an advanced technology to convert heat energy from tension into usable energy if the tether is not stretched then it is perpetual, but depends on gravity

EVEN BASIC GRAVITY MOTORS MAY REQUIRE SOME SORT OF ADVANCED TECHNOLOGY, SUCH AS NANO-ROPES, TO WORK

PMM CONCEPT "GRAVITY CLOCK" INCORPORATING A HORIZONTAL DISCUS

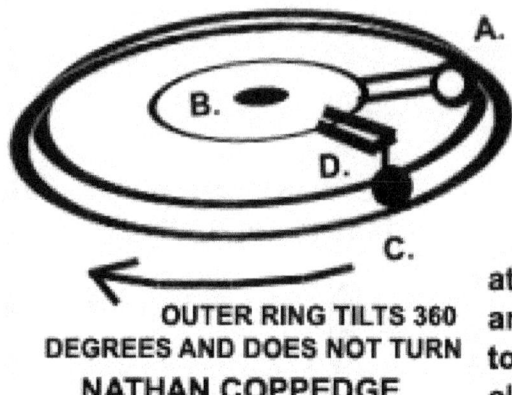

OUTER RING TILTS 360
DEGREES AND DOES NOT TURN
NATHAN COPPEDGE

Ball bearing (A) creates slope which transfers through central lightweight disk (B) to mobile ball-weight (C) attached by member and tether (D), aimed to create in this case clockwise motion

PMM CONCEPT "GRAVITY CLOCK" INCORPORATING A HORIZONTAL DISCUS : VARIATION ONE

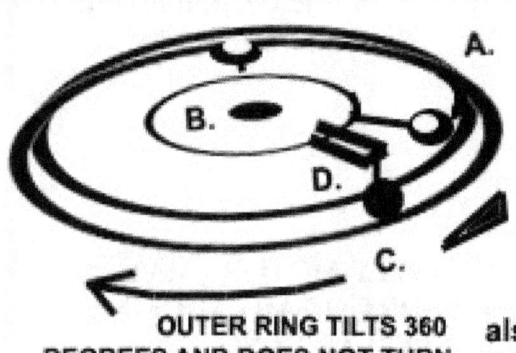

OUTER RING TILTS 360
DEGREES AND DOES NOT TURN
NATHAN COPPEDGE

A design similar the earlier form of gravity clock except involving a 'roller-skate' double-ball-bearing implementation also using a single rolling ball weight for motion

GRAVITY-CLOCK GRAVITY MOTOR CONCEPT. SIMILAR DESIGNS HAVE FAILED TO PROVIDE ENOUGH ENERGY FOR MOTION, E.G. TILT MOTOR EXPERIMENT 2 AND 3.

SPIRAL WHEEL

VERTICAL WHEEL USING SPIRALS AND DOUBLE-DIFFERENCE

A vertical wheel made of spirals and joined from the exterior (not pictured), using two fixed horizontally rotating pendulums; The first [A] is heavier, acting on the second [B], which has enough weight to resist the wheel, providing upwards force against the first pendulum; The rotary motion of the first pendulum is meant to counteract the second, creating a circular motion of the wheel

[SIDE VIEW]

[FRONT VIEW]

Nathan Coppedge

**VERTICAL WHEEL USING PENDU-
LUMS IN AN EFFORT TO CREATE
COUNTERVAILING SPIRALS**

PENDULUM DEVICES

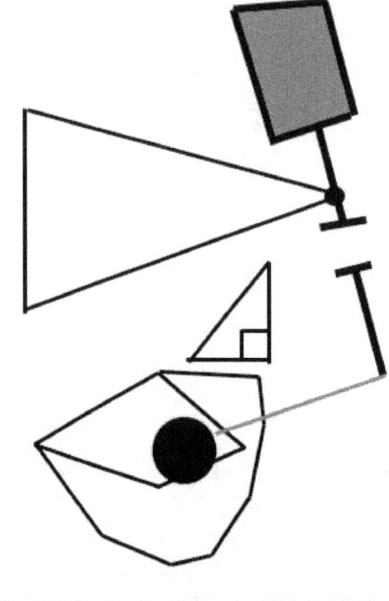

Nathan Coppedge

DIFFERENCE PENDULUM DEVICE USING SPIRAL SLOPE AND STEEP OFFSET DROP POINT AND RETURN DRAG

At drop point (A) mobile ball weight has advantage against counterweight (B); However, according to design when ball weight reaches point (C) it is supported by a spiral track, decreasing effective weight, and causing counterweight (B) to act by dragging the weight upwards; By outwards angularity of the track and upwards pull, the ball weight is thrown in a spiral, until slope allows its full weight to apply at (C)::

PENDULUM TYPE 1

HORIZONTAL SPIRAL DEVICE
USING A DUAL-WEIGHTED CORD

Position A. Outer ball weight follows the inward curve of the spiral, which is steeper than the upward lateral curve supporting the weight; Position B. Outer ball weight continues to follow the curve, as the central fulcrum swivels the counter-weight; Position C. The outer ball weight follows an incline which is more vertical than the sum of the lateral and spiral, so that the weight returns to the start point::

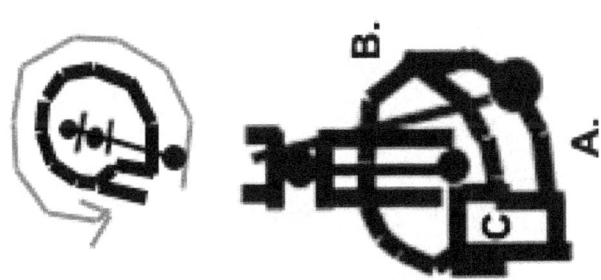

PENDULUM TYPE 2

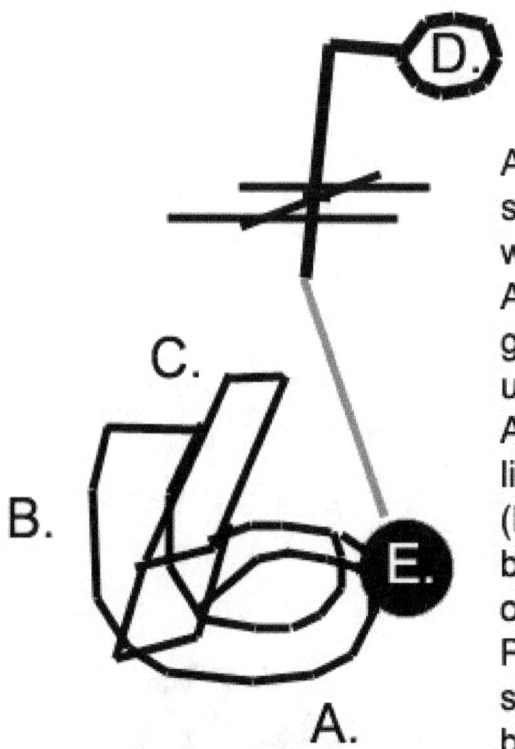

Counterweighted Pendulum, Type 3

At (A) downwards slope pushes ball weight E. outwards, At (B) slope becomes gentle, allowing an upwards climb. At (C) ball drops, lifting counterweight (D.) and pushing ball weight (E.) outwards. Proportionality seems possible because ball weight (E.) is heavier than counterweight (D.)

horizontal support allows (D.) to lift (E.) ⬅ counterweight (D.)

PENDULUM TYPE 3

SPINNING TOP DEVICES

PERPETUAL MOTION "SPINNING TOP"-TYPE CONCEPT #1

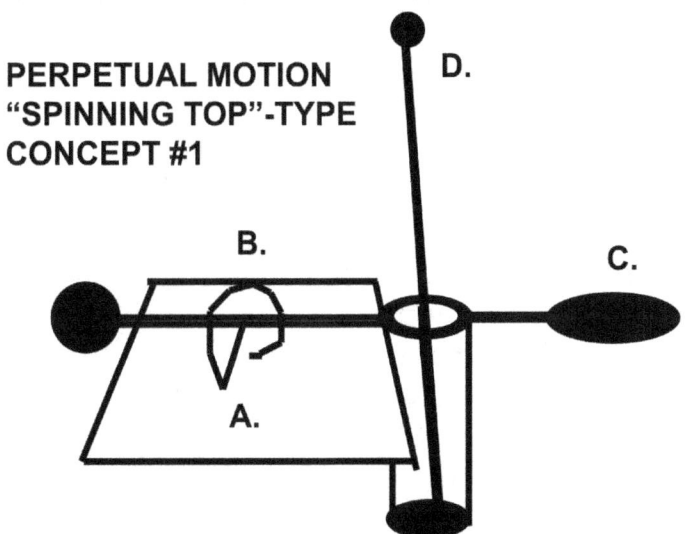

At point A. inward-curving spiral member causes main shaft to rise upon a slope. Main shaft loses altitude but spiral gains altitude. It is eased by counterweight C. The ramp may tilt, allowing the main shaft to swing back to position A. In general, the motion allows pin D. to spin, generating energy.

SPINNING TOP #1

PERPETUAL SPINNING TOP TYPE #2

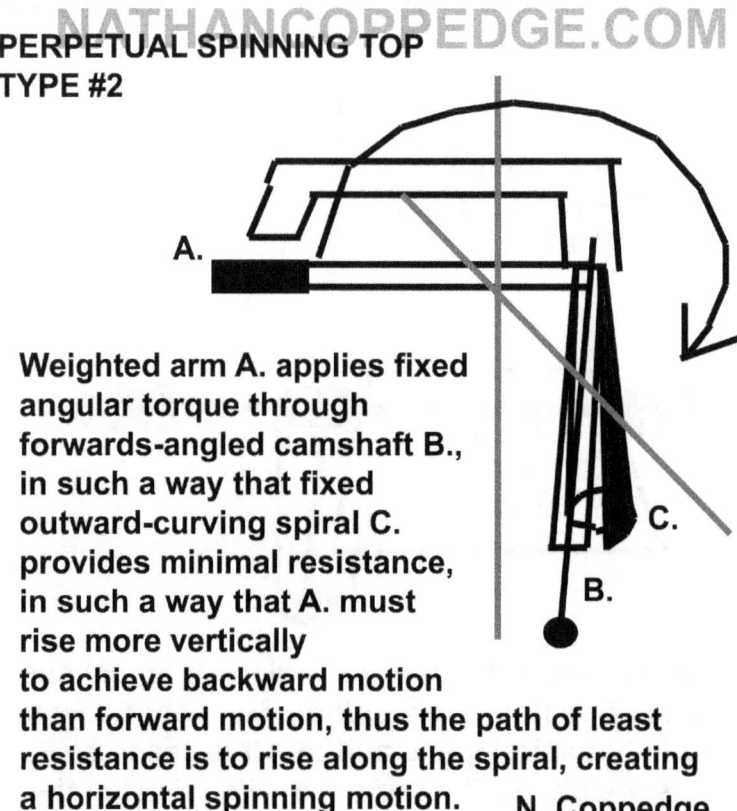

A.

Weighted arm A. applies fixed angular torque through forwards-angled camshaft B., in such a way that fixed outward-curving spiral C. provides minimal resistance, in such a way that A. must rise more vertically to achieve backward motion than forward motion, thus the path of least resistance is to rise along the spiral, creating a horizontal spinning motion.　N. Coppedge

C.

B.

SPINNING TOP #2

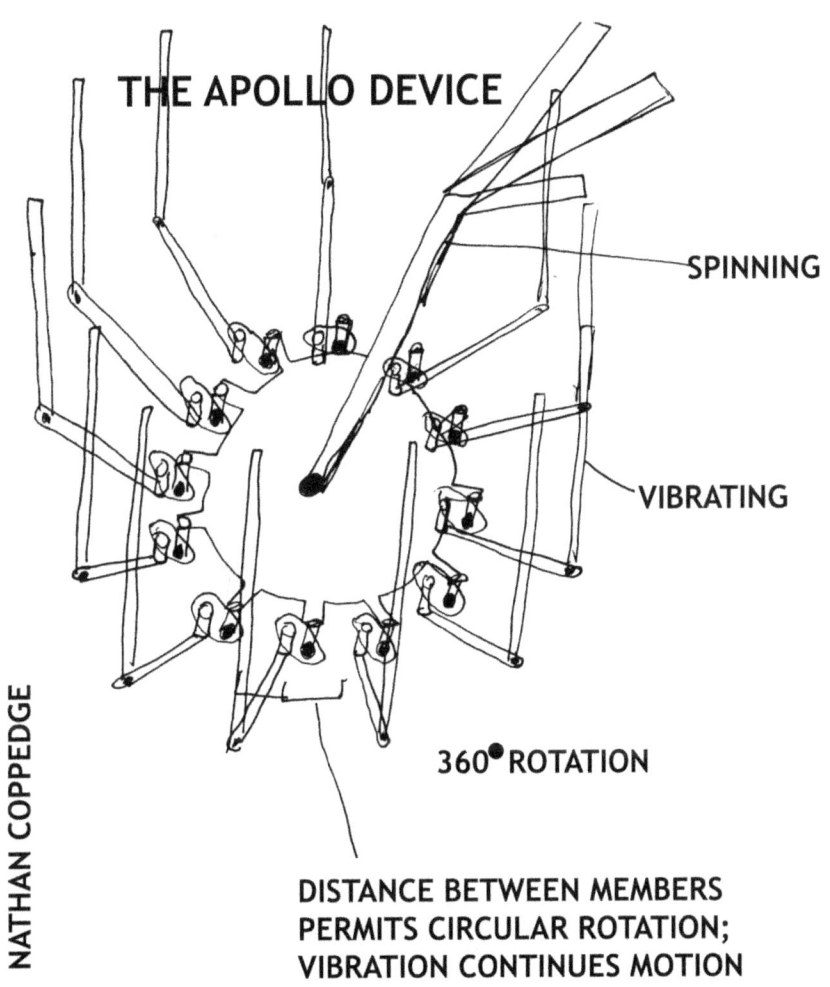

THE APOLLO DEVICE

SPINNING

VIBRATING

360° ROTATION

DISTANCE BETWEEN MEMBERS
PERMITS CIRCULAR ROTATION;
VIBRATION CONTINUES MOTION

NATHAN COPPEDGE

APOLLO DEVICE -- Perhaps existed
in Ancient Greece.

Nathan Coppedge

ODDITY #1

Perpetual Motion Concept Incorporating a Tilting Structure with a Central Arc or Basin and two Sloped Troughs

ball weights are intended to move (1) on an outer bow shape across the trancept / pivot point to a point of backwards slope towards the center and (2) in a reactive position in the center creating greater variance. The intended result is a repeated pivoting motion

5/14/2008

CENTRALIZING

DECENTRALIZING

|| Eucaleh Terrapin ||

PERPETUAL MOTION BALANCE [?]

ODDITY # 2:

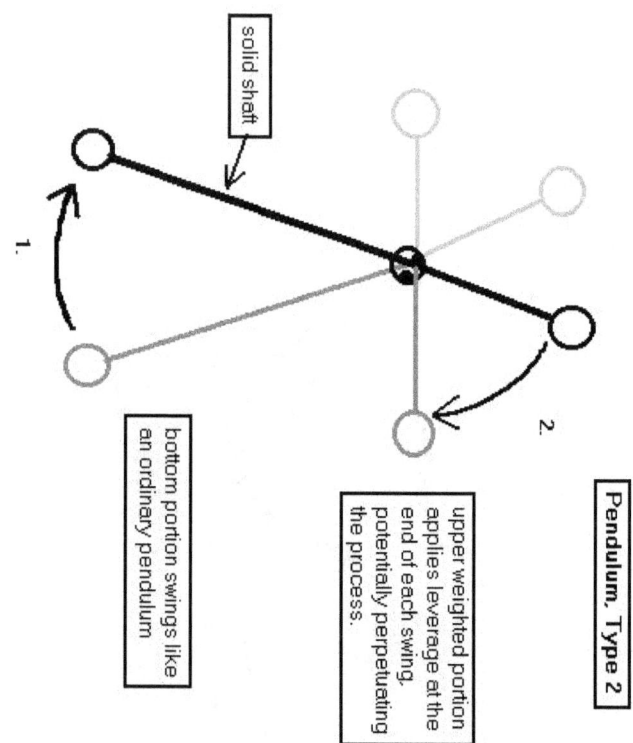

solid shaft

1.

2.

Pendulum, Type 2

bottom portion swings like an ordinary pendulum

upper weighted portion applies leverage at the end of each swing, potentially perpetuating the process.

PENDULUM TYPE 2 [?]:
UPPER PIECE MOVES L/R SLIGHTLY

ODDITY #3

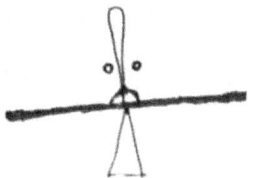

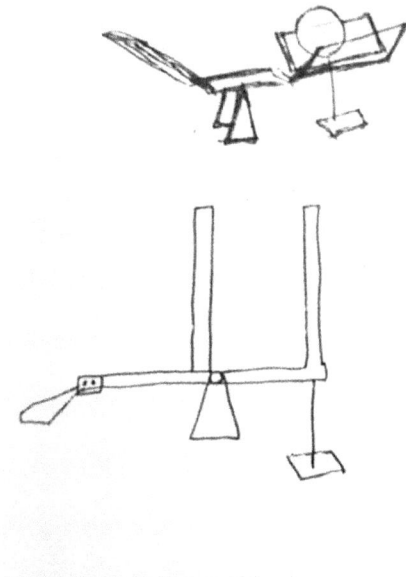

**DEVICES FOR GENERATING
ENERGY IN PUBLIC PLACES**

ODDITY #5

Oddity Using Differential Equilibrium

Theoretically in a divice using the same parts as above, momentum contributes to four "shift" promoting continuous motion, according to a principle of equilibrious motion---

[FOR #4 SEE SIDEWAYS LEVERAGE UNDER REPEATING LEVERAGE]

[SIC] DEVICE.

A DEVICE MAKING USE OF EQUILIB-RIUM (PERHAPS SEVERAL DESIGNS)

ODDITY #6

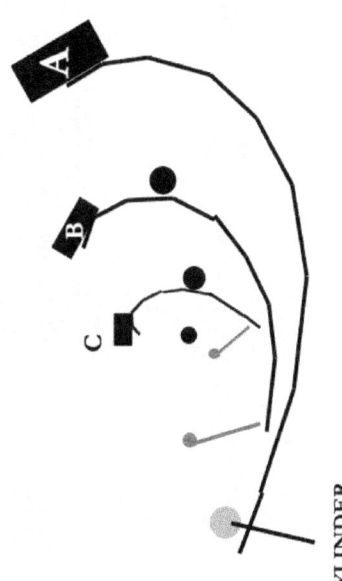

PARTIAL ENERGY CONCEPT
[AUGMENTED ENERGY CONCEPT]
USING TWO-DIMENSIONAL BASINS
IN RECESSED CONFIGURATION

Cylinder enters basin A
with no energy input;
Rolls to tip of cylinder B
at which point A begins
to counteract with
counterweight force;
Cylinder rolls to tip of
cylinder C, activating
basin B counterweight,
providing lift; C begins
to operate, providing
additional lift;

Nathan L. Coppedge

CYLINDER

DEVICE FOR ROLLING UPWARDS

"APERRATURES":

**APPLICATION OF PERPETUAL MO-
TION TO BUILDINGS CONSTRUCTION**

volitional energy gazebo

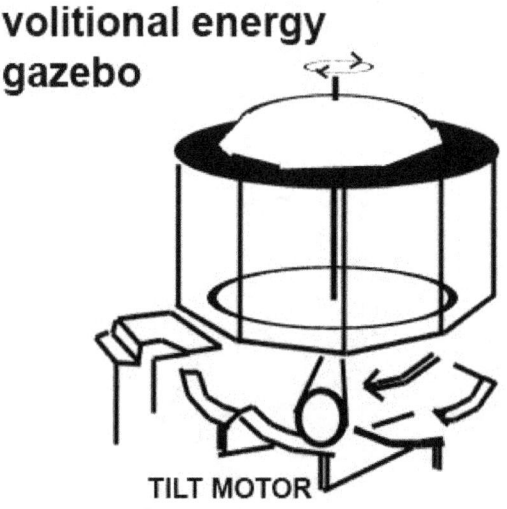

TILT MOTOR

Perhaps the simplest type
of aperrature (mobile building)

PERPETUAL MOTION GAZEBO

APPERATURE RAMP (AUTOMATED RAMP) USING REPEATED VESCENSION

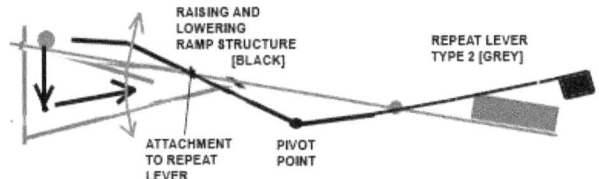

Implementation that would ostensibly allow automated ascension, via a principle of lever advantage versus counterweighted mass and momentum to overcome equilibrium, essentially a repeating lever for buildings; Use is made of a counterbalance, as with elevators

PEPRETUAL MOTION RAMP

Nathan Coppedge

Aperrature (Mobile Building) Using Counter-
Weighted, Ratcheted Tilt Motor

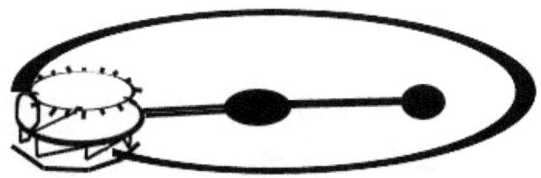

Here a free-floating Tilt Motor apparatus,
augmented by counter-weight, operates
a ratchet, creating a circular movement

PERPETUAL MOTION RATCHET

APPENDIXES

ADDITIONAL LAWS AND PRINCIPLES

TRINITY OF PERPETUAL MOTION

 I. (a) Total H + V support = 90° at every point, (b) Average altitude is constant,

 II. (a) Momentum without velocity, (b) Unbalanced principle,

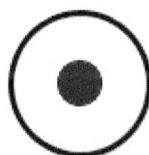 III. (a) Cyclic (b) Mechanical

APPENDIX/

FOR SIMPLICITY'S SAKE

Traditional View of Machines
1. How it's made.
2. What it's for.
3. How it works.

It may be simpler to say that there are two senses of the machine:

1. What it repre-  is. Linear thus quantitative, sented by the symbol:

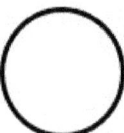 1. What it does. Typically a spiral

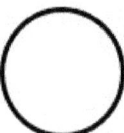 Modified by time to equal a circle:

There is a still simpler case, which is what we have to look out for. The simpler case is simply 'IT IS *SOMETHING*'

Not only does this final view imply common-

place assumptions (the bane of the true philosopher), but under this view one assumes that
things of one kind always works one way,
which may not after all be the case.

Returning to the three categories we need to
prove (reverse of previous): 1. That it works, 2.
that it serves a purpose, and 3. That it is not
purely theoretical.

So, we begin with theory, we apply purpose,
and if the two are flawless, we end up with #1.

It is easier to picture how it might work if we
understand that 'what it is' supplements, but
does not necessitate what it does (contrary to
the conventional view).

We know that if these machines exist, they are
physical, and they are machines, and so they
do not automatically work unless they are in
precisely the correct configuration: they CAN
be stopped.

And hopefully started again…

Although starting should involve no more than
a relationship of moving parts.

APPENDIX/

APPLIED PRINCIPLES OF PERPETUAL
MOTION

Escher Machine

Principle: H - V > V(H). Referring to angularity.

Proof: H - V > V(H).

Modular Trough Leverage Device

Principle: Supported Leverage

Proof: [1] Individual units have been proven to
go up, down, then up again, without using
power, and return to their initial altitudes, [2]
The repetition of the units requires no additional principle.

Tilt Motor---or "God's Rolling Pin"

Principle: Momentum Without Velocity

Proof: [1] A person applying leverage against
his or her own weight can lift him or herself
smaller distances than the motion of the lever.
When this motion is applied to changes in
slope, a very small change of distance is neces-

sary for horizontal motion. [2] Because the wheel is horizontal, only slope is required to extend motion. [3] The motion of the levers is greater than the motion to create slope because the rolling cone is tall enough to do so. [4] If resistance is minimal, the result is motion.

Repeating Leverage

Principle: A counterweight may be operated by leverage. The counter-weight applies constant force, which acts on the lever whether or not it is activated. Thus the lever may be activated repeatedly.

Proof: [1] Slope is possible continuously when upwards and downwards movements are permitted. [2] All that is required for an upwards or downwards movement is a shift in mass; For upwards movement, advantage is required, for downwards movement a disadvantage is required. Thus the system is balanced in force, but unbalanced in momentum. [3] A differing position is acquired by acting upon a mobile ball-weight, which moves because continuous slope is possible.

Motive Mass Machine

Principle: A domino principle in which the dominoes do not need to be reset.

Proof: [1] A weight in free-fall can move an equal weight horizontally when the second weight is supported on wheels. [2] A chain re-action of this type would gain momentum, because there is a principle of advantage, much like dominoes. [3] In this case, the average altitude remains constant, producing an over-unity principle, because the 'dominoes' 'do not need to be reset': they are re-set, but only by the principle of advantage made possible by the vertical versus horizontal rule, e.g. because part of the weight is unsupported at a given time.

Mass-Modular Devices: Grav-Buoy 2 Curving Rail Device

Principle: Advantage is had by multiple units, against which is placed minimal resistance.

Proof: [1] In the final design of the Grav-Buoy, the weight of water is cleverly reduced, making the example an exceptional case, [2] In the case of the curving rail, perhaps an absence of friction would make a similar cycle chain-reacting, if not mobile; This has an advantage against some concepts of the Bhaskara Wheel, because the upwardly mobile portion is supported by a track. This is made possible by a flexible chain-link or cable structure. It can be seen for example, that a car on wheels is much easier to push than a car without wheels.

APPENDIX/

RED LETTER DAYS OF
PERPETUAL MOTION

JULY 3RD, 2014
Day I proved an object could roll upwards
of its own volition (attempt to build the
Escher Machine).

NOV 10, 2013
My First Successful Over-Unity Experiment

If a device can go up and then down from a
position of rest, and all parts return to their
original positions, then there is theoretical
over- unity. This device meets the criteria.

The diagram shows how to turn the over-
unity device into a perpetual motion machine.

APRIL 26TH, 2009
Day I Invented 50 Devices

These devices contributed significantly to
my working tools, but so far as I know none
of them have joined my online collection. I
cite many flaws.

APRIL 16TH
Day in which I have invented Five Devices
more than once, first in 2007. One of these
was the Coquette.

APRIL 2ND, 2007
Day I found evidence that the Tilt Motor
could work . With near horizontal slope,
leverage can extend slope. Confirmed once,
denied twice. Confirmation was with a level.

FEBRUARY 4TH, 2007
Day I seemed to prove an unbalanced wheel,
the Principled Asymmetry. The device
seems to rotate more easily in one direction
than another.

OCTOBER 30TH, 2006
Day I invented the Tilt Motor, and time-
traveled back to the morning of the same
day The Tilt Motor is the ingenious concept
of a horizontal wheel, operated by levers.

APPENDIX/

Solving a Flummox

You've seen a design operate significantly at least one time---then the apparatus
begins to fail to work. What could cause such a
problem? It's called a flummox.

A flummox is a quasi-magical problem
that occurs when the powers that be determine
that something is in too much doubt to reflect
reality. When this happens, even very basic
parts cease to work. There is a literal feeling of
finding 'gum in the works.' This gum may be
invisible, but the powers that be insist that it is
there. You can feel them using phony proofs,
like 'tape is sticky' and 'it takes a trick,' etc.
(However, it does not really take a trick in the
absolute sense, at all, unless the principle of
combining principles is somehow inauthentic
to reality, which it isn't. In fact, combinations
make reality much more authentic than it
would otherwise be. Complexity is what makes
engineering so adequate).

A flummox can traditionally be solved
only by the most deliberate work. But another
method is to convince a professional manufac-
turer that the design works as a theory. A still
further method is to demonstrate the design in
stages---each one of which may serve as proof-
of-concept. My Successful Over-Unity Experiment 1 is an example of this.

111

AUTOBIOGRAPHY OF THE INVENTOR OF PERPETUAL MOTION

By Nathan Coppedge, Inventor, Theorist, but Not Builder

INDEX

Nathan Coppedge

RHETORICAL INTRODUC-
TION:

1. Finsrud's Perpetual Motion Sculpture dates from a considerably earlier period than my designs. But, since he has not written an autobiography, I will consider his claim as false.

2. Many inventors have been discredited.

3. Electricity-based engines clearly consume fuel unless the electric component is merely a generator added on to a machine that already operates without electricity.

4. Simple over-unity is the only option.

5. There are only so many theories viable in this category, unless you believe in infinite possibility.

6. In my life, I have found some evidence of over-unity. The cases I provide are sometimes simpler than Finsrud's example, particularly if he is a fake using batteries.

7. I'm not trying to ask trick questions.

8. Instead, I'm trying to convince you of actual evidence.

9. Actual evidence is hard to believe sometimes.

10. I know if you had good intentions, and omniscience, you'd go straight to my diagrams.

11. Otherwise, the only pretense is amusement.

12. But, perhaps you are interested in a good story. So that is what I am providing. This is the story of my life. Although perpetual motion is a theme, it is not the only theme of my life. People should avoid thinking that I have to be 100% made of perpetual motion if I am an inventor. I am still a man. I am not a robot. Essentially, I am an inventor---maybe not the only inventor, but at least someone with a remarkable claim.

MY STORY

On the Ivy-Covered Hill Before the World Began

We were at a crossroads, about to depart for my father's research trip in a foreign country. I knew some Spanish at this time, because my father always had it in mind that we were going to go on this trip. We were friends with the H's next door, who were an intellectual Jewish family (their father would later work for the Mars Corporation), so from the very beginning I had the feeling of being quite international.

The house we lived in was a duplex, with the H's living next door to our right. Being in the vicinity of Yale, New Haven was such a town as to have annual tag sales from all the students moving out, not just around campus, but also in the Orange St. neighborhood where my mother later lived with us.

Some of my first experiences involved a tag sale with a rocking horse. For those that don't know, a rocking horse is a kind of plastic or wooden horse arrangement that swings on ropes or springs attached to some sort of base structure. This particular one was plastic, but its springs were metal and had no covering. I remember having a deep love-hate relationship for this horse.

Foolishly, I would test the horse again and again, wanting to have fun. But

121

every time, I would end up with a new pinch on one of my fingers, as I tried to stop myself from falling out of the seat.

Eventually I decided ---perhaps based on my father's advice---to not ride it anymore, but just to watch the horse and admire it. I remember one time when it was very late, my father called me to come inside, and it was already getting dark.

It was one of my great early learning experiences, to respect the pain that the horse commanded, while also not holding the horse entirely responsible. Perhaps out of my sadness for leaving home, I thought of the horse as if it were a real person, with real human flaws. The horse was my spiritually imaginary addition to our intellectual family.

A Day of Dance for the Dead

From a very early age, I was filled with intense emotions. I was empathetic with the deepest feelings people had, but I also understood that many people were liars who did not want to share their feelings. The world was consequently full of 'dead people' to me: people who would not bare their souls. Then, it was something of a coincidence that, at the age of three years old, I found myself, due to my father's graduate studies, positioned in Caracas, Venezuela during the Day of the Dead celebrations.

These festivals take place throughout Central and Latin America, and involve celebrating the lives of the dead ancestors. It became the first, most vivid symbol for me, not just of death and re-birth, but of a 'mechanic of bones' which underlies all things. Seeing images of industrialists depicted as skeletons, I could think of Death himself as a kind of over-mensche thread weaver. People who thrived in this world were people who knew about machines, I thought. All of this coming from a very disillusioned little boy.

I remember the specific moment in which I had this realization. My father was buying an enormous lollipop from a hand-cart during the festival. Apparently, he was trying to sweeten the deal. Having this thought was the nearest thing I could be to a business man, amongst all these people in fancy costumes.

Once, in Venezuela, I nearly lost my father, during an occasion at a public swimming pool. After that, I started to think that part of growing up was knowing how to make diabolical bargains.

The Eye of the Beholder

I remember the next part of my life as a succession of houses, escaping the Spanish language. I had sworn when we left Venezuela that I would never speak Spanish again. It wasn't until I took up Spanish language in middle school that I encountered it first hand again.

The first house I remember was somewhere in New Haven, a temporary arrangement that happened while my parents were still married.

On one particular day, my father had snookered my mother into watching a terrifying horror movie. He told me again and again 'stay in bed'. 'Stay in bed'. Which it turned out was only a sleeping bag in the next room.

Then, drawn by the ominous sounds of the television, I crept up behind the couch.

There, in the middle of the television, I could see one livid eye, staring at me like it was consciously alive!

Then, what was more terrifying, my father turned around and saw me. He was staring fearfully, his face completely white, like he had seen the most terrifying thing----and it was me. The revelation that he was afraid made me even more afraid. "Get to bed!

125

Nathan. Get to bed!" he said. It was probably in that moment that I had my first feeling of de-personalization that would later mark my mental illness.

What was more frightening than the movie to me was that we were in a cold house, and all I had to sleep in was a dry old sleeping bag.

Rainbows

Things became more difficult when it was discovered that my younger brother was a child genius. He went to a daycare that looked like it was cooked up by Lucifer, and for that reason I didn't feel like attending it myself. They had really amazing songs which filled me with hazy emotional marvel. My mind was pre-occupied with how my brother loved the encyclopedia more than I did.

Life was simple. My mother was independent, and working at an environmental agency doing office work. It was a dream job for her. We would get mailings related to environmental news, including some crude comic books showing vivid caricatures of fossil-fuel polluters and mad scientists, confronted by brave, sagacious environmental superheroes. We also went to a food co-operative, where my mother could pay for bulk foods like rice cakes and healthy-brand carbonized beverages.

At this time we were the kind of family that was proud not to eat fast-food. Although we sometimes ate meat (except for my brother, who is a vegetarian to this day), we ate pasta for dinner almost every day of the week. This gave me plenty of mental fuel, but not much substance to go on.

Me and my brother were both thin, and both of us considered this to be the best and healthiest way. Neither of considered it to be unhealthy. And we always got helpings of dessert at our grandmothers' and candy on Halloween, so it seemed like the natural way. It's not that we didn't eat. We had breakfast, lunch and dinner, but it was the 1980's, and thin was okay.

My First Kiss

I had a fondness for a girl named "Emily" in elementary school. We exchanged valentines and she thought of me as kind of a naughty boy. She was allowed to visit, and we were told "not to do too much kissing".

This got the female to thinking that kissing was "necessary". So she waited in a big drawn-up room with the idea that I was going to come and kiss her. And this made me think of my big medieval fantasies about how kissing is the be-all-end-all of love.

I was partly worried that it meant I'd have to marry her. And I was worried that kissing might even kill me. They were similar words, after all.

So, then I went into the room, and said: "I won't kiss you on the lips, I'll kiss you on your arm".

So, she leant her arm, and there I planted a kiss.

"We kissed" she said. And for a moment, we had the feeling of being married together.

"Now we have to make house" she said. But I was already becoming disinter-

129

ested.

Years later, my mother had me talk to her on the phone, and I couldn't even understand her accent, because she had moved to Colorado. Apparently, the consummation was everything for me, which is exactly what her mother said when we had to talk about it.

Nonetheless, it seemed highly romantic. All the more so, because I later learned that she had the name of a newscaster who had taken her clothes off.

Quest for Immortality

Once, when I was nine years old, I decided that I was done playing death. I had finally abandoned all of the Spanish language. I was like an English king----a Justisivus Rex---and the Latin language deserved a plan of action. I decided---critically, and emotionally----that I was no longer death, I was merely *old*. I had always been old. I had already acquired every wisdom that I would attain.

I had, according to my view, every secret I would ever want, the kernel-seeds of every symbol I would ever want to un-chamber, already written into the source of my life. I had, for instance, a library like Babel, towering in the middle of Yale. I imagined that at some point (as I later did briefly), gain access to such a library. The library contained books, but there were also stones, and water, and tea, and laughter which would entertain me for eons. In my reasoning, like this, my logic was that I was old before I was young.

As I later worked out in clearer terms, I had sacrificed my immortal youth for age, followed by pragmatic youth. I would not know immediately whether my plan succeeded. After all, time's eddies are unpredictable. America's law might be perversity. I might be opposed by evil scientists, or diabolical incarnations. But in a place as happy as New Haven, the trouble could not be very significant to an immortal. By protracting the longest life, I meant to live the optimal life. Perhaps I would even live immortally. That was my idea.

Throughout my years in the public schools, I kept this thought in the back of my mind, as an unspoken, utterly silent idea: "I am old before I am young". Indeed, I remained strangely un-athletic, and people remarked that I was wise.

Ivanhoe, Ivanhoe

One of my most turn-around experiences happened as I was reading the novel Ivanhoe in 5th or 6th grade. I prided myself on this, and probably only thought it was manageable because I already thought I was immortal. The language was troublesome in places (this was one of the earlier editions), because of the ancient Anglo-Saxon language, but I deliberately read pages and pages, until, after a number of weeks, I reached somewhere near the midpoint of the book. Bear in mind that my eyes were not in great shape----I have always suffered from a condition called astigmatism which makes it difficult to read anything. But, like Wilson's battle with Polio, I took the challenge as an opportunity to become absolutely obsessed with reading. In any case, although I never finished Ivanhoe, or at least never memorized the plot, I considered it to be an achievement that I made it that far. I remember vividly the scene in which a dog blocks the path of the Saxon, and, he, fearful of being stricken with bad luck, hurls a spear, and kills it. Although I struggled with this motif, I tried to translate it as something that would mean something in the context of immortality. Perhaps I could be lucky, or perhaps I could find a 'way across a dangerous path'. In any case, it would be better than fighting in a tournament.

Puberty

At the age of 12, I had been told that when I turned 13 I would probably start to masturbate. So, on the day I turned 13, I waited as long as possible. It was about 11 o-clock in the morning.

I seized my penis, and began going into spasms. Then I put my penis back into my underwear, but I felt an ecstatic rush like I was the head Indian on a course of rhinos, and leaned, with the purest feeling of total ecstasy, with my pants wet and open, into the only thing I could think to use, which was a jagged sheet of note-book paper.

In that first experience, at least three broad rivers sang my body electric. I left blood and semen marking some kind of arcane pattern on the lined sheet.

At first, I didn't even feel guilty. I felt restricted, but also brave and free. In the years that followed, I repeated the act over and over, many times returning to the notebook paper.

The paper gouged into my skin, creating a cleft, which wore away the sweet spot under my flesh. In the moment I became a passionate man (or some say, less than an animal), I also became an intellectual by disposition.

From then on---what I called Maitre d' Papel----I had a passion for paper, not just in the privacy of my room, but also for writing and drawing on.

As the years went by, I learned to touch myself with my hand instead, but the passion I learned with those pillows and sheets of paper never seemed to quite return in the same way. The passion was there, however, for many years, until I was forced to take a medication which had a side effect of reducing my 'libido' as I was forced to call it.

Kit & Company

Puberty was sort of a blur. I did school work, but in private I was hooked on masturbation. Sometimes I would masturbate two or even---rarely---three times in a day. The first significant thing that happened was what I took to be love. It was not mutual love, but it was total, chivalric, love-at-a-distance.

I didn't have the language powers to communicate my feelings. I think I was still developing the idea, based on my programmer brother, that physicists were the ultimate lovers, because they knew how to look so nerdy and 'in love'. Come to think of it, that was exactly what was happening.

It took place at a Science Fair exhibit. I was in my first year of high school. "Kit" was in my brother's class, still in her last year of middle school. People might say she had over-developed breasts. I would have resented that statement entirely.

The curves on this woman were out of the ordinary. She had hips like a milking woman or Dutch goddess. Her breasts----I called them tits and then took it back---- were darting, round, and still growing.

She was the woman that everyone lusted after at the Arts School later on. I didn't overcome my lust for this woman for about a decade, maybe 15 years later. Some men looked on the fact that she later got a breast reduction, and just got more horny. But it wasn't because of the small size of her tits.

My memory of "Kit" from the Science Fair was like an impressionist painting in the back of my mind, fomenting dreams and making everything seem better than it was.

Out of Unity

At the end of high school I was undergoing a transition phase, as I coped with the idea of how to sustain my intellectual ambitions, which were growing strange and gnarly.

I had never been an expert at French, or spoken very many proverbs. My study of poetry in high school had been an academic success, but in retrospect looked like a desperate attempt to get attention from girls.

Instead of turning around my mentality, I decided to go forwards with the same stuff in my first year in college. I couldn't risk changing my entire attitude in just one year.

Meanwhile, my attitude was changing, in a sort of unhealthy way. I stopped doing homework for my last English class, and on Sundays I would bomb at my friend "Garth's" house borrowing time to play video games on his superior P.C. computer. The video games distracted me from schoolwork, both at "Garth's" and at home.

The past several years I had been going to a Unitarian Universalist church on Sundays, which Garth also attended. There were a series of climactic events at the Church, the main one being a time that the Sunday class in the Carriage House decided on its own to take an adventure through the streets in the wealthy neighborhood behind the church. At that point, (I determined this myself), my mind had a fragmented feeling of being both tempted by the sinful wealth of the neighborhood, and not being entirely sure that I was safe. This also would lead to my later sense of my mental illness.

The Unitarian Church was also a symbolic place, due to the wooden carvings in the main room. I could interpret from these a sense of categorical organization which would assist me in regaining philosophical, if not psychological, rationality. The Church also accepted some of my writings to keep on file out of friendship.

Escaping Onion Flower

Bard College, 2001: I had just come back from my father's vacation in Great Britain, where I made some sketches of buildings, and thought about Romanticism. I was moving in to a liberal arts college, in what were called 'Toasters' --- which were a particular type of modern or post-modernist dormitory.

My roommate was much more musical than myself, and had decided to play loud ballads by Nirvana and rock & blues bands. One of the songs sounded Russian, and featured destructive noises. I couldn't bear the music, and broke down in tears, what I thought to be a great act of empathy, thinking I could somehow save Russia from a nuclear war. My American arrogance said that Russia must be the victim, although I could stand to feel sorry that they might suffer destruction. Saving Russia, I wanted to save the whole world.

With this mentality going on, and beginning to become disorganized about my classes, I first heard, when I woke up around 2pm, the news of the September 11th attacks. At first I thought that someone had dropped a bomb on New York. Relieved that it was just terrorism, I felt major endorphins, like I was an emotional superhero walking amongst emotional zombies. When an ambulance offered to take student volunteers, I continued to think that maybe we were on the brink of a nuclear disaster.

I began to hear voices in my head: something I hadn't heard of before. The voices seemed like spiritual authorities, although they often said mean and distasteful things, commanding me, and giving me hints about what to think. I thought of the voices automatically as my intellectual mentors, but at the same time it was obvious that there was something malicious going on.

Perhaps these voices were a manifestation of a word I did not acquire until later----some sort of 'Lamarckian' program to weed out unintelligent students? If so, I felt like I was being unfairly treated. After all, people like "Kit" wouldn't find this kind of treatment. They were desirable to all. If "Kit" were smarter than me, it was a cruel, unbalanced coincidence. But I couldn't help but worship her. So, maybe it was that way for others, too. Maybe, I realized finally, she was an actual goddess.

I felt distant that Christmas. I learned that I got A's in all my classes----I made the Dean's List for Bard College that semester. Things seemed uneventful, apart from my new, weird reality. I tried to conceal my voices from my mother. Retrospectively, she said there must have been something going wrong. But then, in January 2002, I had a brief schizo-manic episode, where I wrote down many arcane-sounding thoughts. Suddenly, I couldn't sleep. I called medical staff three nights in a row, feeling a sense of impending panic that I couldn't control. Then, around the morning of the 12th of January, I had a hallucination of a square sun----not blotted black, but instead, glowing in a square shape. As I was escorted to St. Joseph's hospital by a driver, I had further visual hallucinations: first, of a terrifying transparent bicyclist 'ghost', and then of a similarly transparent squirmy creature that greeted me (without any gesture) when I arrived at the hospital.

At the hospital, I had the nearest thing to a panic attack that I have ever had in my life, a kind of irrational stage that was frightening and difficult to bear. I felt the roots of reality dissipating, and noises crept in like the sky was peeling with a gigantic lawnmower. What must have been a skateboarder, but which seemed at the time completely unexplainable. It's possible the skateboarder was not even in that area of the city. I was afraid my hallucinations would continue, but was adamantly against taking any medications.

My mother was supportive of me, and when I called her, she said she would come with an entourage of supporters.

These supporters, for whatever reason, did not seem completely reassuring (I still had the dream of being a macho professor), but at least I had the faintest feeling of love from the world, which was remarkable at the time.

Later I would describe this time---- the notebooks I kept, the sketches I made----as the time of the 'escaping onion flower'. I was an 'onion flower' --- a kind of philosopher who was very tender, who would pay off nicely.

Last Mandalas (Not Dead Yet)

It wasn't until 2003 (the first hospitalization occurred in the beginning of 2002), that I was hospitalized again. This time I had impulsively trespassed on the roof of a Yale property---perhaps attempting to re-live a childhood experience, or discover some new 'visionary' reality. I was not on drugs. I was simply in what is called a 'psycho-delusional' state. I thought my visionary view of reality was more important than the laws of physics. And I almost thought I could levitate.

Fortunately, I decided to break a window instead of stepping off the roof. Then I was arrested, and spent the second of two nights in jail, until my mother bailed me out, and we hired an expensive lawyer to explain how I was psycho-delusional.

A deal was arranged where I had a choice: I could go to the hospital to get treated for schizophrenia, or I could go directly to jail. So I decided to be hospitalized.

Then I entered the time I call 'my last mandalas'. I followed a pattern on the floor with my footsteps, eking desperately after my own authenticity: a time when I was not threatened by medication. I had not yet been forced under a needle, or

145

given any pills. I had rejected the anti-depressant offered to me at St. Joseph's.

When I made a performance of standing on one of the air ducts next to a window, word was given from the clinician that I was supposed to "take a medication." So I thought fast, and mentioned my "academic parents". And that I had "probably a high I.Q. which shouldn't be damaged". I had heard horror stories about Haldol and how it has nasty side effects. So they hooked me up with a fancy new medication called Respirdal. Which sounds scary, but I was relieved at this point to know it was better.

Incidentally, this hospital offered a class (finally, I thought), on drawing man-dalas, which was the kind of childish thing you offer to potentially creative mental pa-tients to give them a sense of hope. Ironi-cally, I felt that this was the last time I would authentically deal with them.

When I got back from the hospital to my mother's house, I was nervous about the side effects, because I thought all of the medications have to be pretty bad, or none of them would have bad side effects. There were some symptoms of rashes on my legs, and bloated legs, and once I was forced to walk oddly. That's when I real-ized my last mandalas were over. God, somehow, had lost his standard of reality.

Theories of Sense

I still felt like something of a poet when I vacationed with my Dad at his home in Indiana, in 2005. He was nice enough to take me to the local library, where I resourcefully found a book by the modern poet Theodore Roethke on his theory of poetry.

I became inspired to submit poems to the New Yorker --- they were never accepted ---- but I entered a new poetic phase, in which I wrote what I would later call 'the sublimist poems'. I had written poems before on a Nietzschean theme. These new poems were more lyrical, relying on a mad kind of sense.

When I felt acquainted with poems, I also felt acquainted with a larger part of life. I felt I now had the authority to do something significant --- because I could say it in words. In short, I became more ambitious.

Dot Calm

I founded a website in 2006 devoted to my new passion for perpetual motion machines.

I had strikingly few experiences with the concept as a child, just a few glances into books on mixed scientific subjects.

In 2005 I had been taking classes at Gateway Community College, and became fascinated with inventing in general. When I was 9, I had thought of my first invention----a long tube with a cigarette lighter I called a 'popcorn gun' ---- it didn't work, but it led me to feel competitive enough to make other designs eventually.

In 2005 I thought of the CatSpur shoe concept, probably already invented during some earlier time----I never knew. It involved shoes with long spring-spurs, which would assist in walking, but was im-practicable on stairs. It would make any-one fall down, unless there were ramps everywhere. Elevators in particular could prove to be a problem.

I also thought of a compensation clothing concept that was later adapted by the Big Idea Group, although without giv-ing me any credit (many of the concepts existed already previously).

149

Nathan Coppedge

My final invention of 2005 was a perpetual motion buoys concept, which it turned out had already been invented by Frank Tatay, as recently as 1929. Now I felt I was on to something.

Prospective Legacy

Between 2006 - 2015 I made fifteen or more major concepts for perpetual motion, with many variations within each type.

Some of these were leverage concepts, or used a supporting ramp to create an un-balance with an equilibrizing element, or used other tricks of proportionality.

 I was not short on poetry to describe what I saw as the real, true theories of over-unity. I described the quest poetically in the following verse:

> "While they were floundering...
> I was pondering...
> No more wandering....
> Through the dark tunnels....
> Of grim determination...
> For no; It is time to grow....
> In a thousand-folded folds....
> For which we need an infinite fuel!"

Evidence Against the Classical Model

Around midnight on November 9th - 10th, 2013 I conducted an experiment with a leverage apparatus I built which seemed to confirm a partial perpetual motion effect that I thought was already over-unity.

According to what I observed in the media, no one had before built a device in which a rolling object triggers the same object that caused it to move to move upwards, and then the object has the force to move upwards again, with no net loss of altitude of any of the parts.

While I tried to demonstrate the device to my relatives some time later, they appeared to be blind to my success. My step father said 'You did it, you built the perfect machine'. But his tone was ironic, he decided, and he immediately took back what he said. He was convinced that the objects lost some altitude from the starting point, which in my view was not true, because the marble could be stopped before losing the altitude it had gained, and the lever could still be triggered within the space in which the marble gained altitude.

That the event was subsequently ignored by the media is an understatement. There is little else to say about it--- until someone accepts the theory.

Nathan Coppedge

The Escher Machine, The Escher Machine

The next year I conducted another experiment.

I was following intuitions about the interaction between track and slope that I observed with the previous example.. My conclusion was that there may be a particular angle in which horizontal slope --- such as the left- or right- angularity of a wall --- might be able to influence and overcome vertical slope.

I called the machine that was a by-product of this reasoning 'The Escher Machine', because it was the nearest thing to creating Escher's illusion in nature.

The theory of the machine is that it is capable of upwards and downwards movement from rest, without expending energy!

In numerous experiments that followed, I found enough evidence to be persuaded that under certain conditions such a device could actually be built!

The Media Plays Like a Drunk Parrot

At this time----with two experiments mostly substantiated and recorded with video evidence, I waited for media attention to pour in. But, so far, in vain! Media appeared completely unaware and oblivious of the remarkable achievements! It is not that this is an appropriate reaction, but that that is just how it happens to be!

I had to tolerate it! With a reaction like this, I suddenly knew that even a working perpetual motion machine would not necessarily grab the attention of the media! And, if that was the case, how was anyone supposed to develop the psychological momentum to build one?

To this day---fortunately not so far later----I await new and improved scientific accounting of the provable theories of perpetual energy that I have already demonstrated and made publicly available.

It may be that I don't have the exact evidence that would animate the media, but at least I have many fascinating diagrams which give evidence of how----for once----a real machine might work!

**NATHAN COPPEDGE,
INVENTOR (2014)**

A PARTIAL OVER-UNITY DEVICE IS SOMETIMES MORE
COMPLICATED, AS THE ABOVE 'PRINCIPLED ASYMMETRY'
- - -

Nathan Coppedge

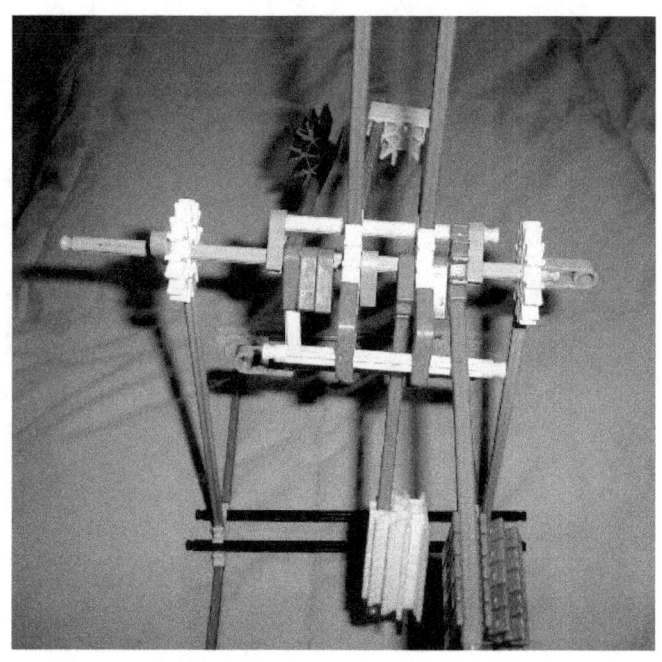

ADDITIONAL DETAIL OF THE 'PRINCIPLED
ASYMMETRY'

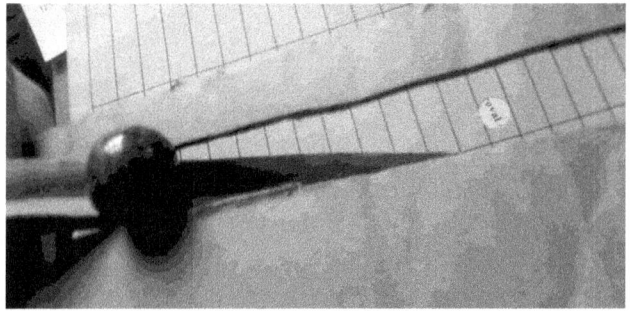

COUNTERWEIGHT (NOT PICTURED, AT LEFT)
MEETS LITTLE RESISTANCE WHEN THE
MARBLE
IS SUPPORTED. INDEED, THE MARBLE CAN
EVEN
BE LIFTED BY ITS OWN WEIGHT EQUIVA-
LENCE.

AFTER-NOTE

To this day (Feb 25, 2015) the legacy of perpetual motion haunts inventors. If there is a viable principle, something should work. But unless experienced builders---not just theorists---encounter the problems of perpetual energy, it is unlikely that all but the most simple and easy-to-build models will ever be attempted.

BOOKS BY NATHAN COPPEDGE

RELATED BOOKS

    THE SCIENTIFIC PAPERS

THE DIMENSIONAL ENCYCLOPEDIA

PERPETUAL MOTION GENIUS'
    GUIDES

PHILOSOPHY / NON-FICTION
    ARCHE-LOGOS
    BASIC PLATONISM
    THE BOOK OF PARADOXES
    COHERENT LOGIC
    HOW TO WRITE APHORISMS
    INTERMEDIATE INSIGHTS
    METAPHYSICAL SEMANTICS
    THE MODIST MANIFESTO
    THE NINESQUARE NOTEBOOK
    SECR. PRINC. OF IMMORTAL.

ART BOOKS
    HIGH ART
    N.C.'S HYPER-CUBISM

POETRY
    CREEPING CADENCE
    THE OLD INCANTATIONS
    POEMS BY GOD

THE DIMENSIONAL ENCYCLOPEDIA

The Dimensional Philosopher's Toolkit
The Dimensional Psychologist's Toolkit

TO BE RELEASED 2015 OR LATER:

The Dimensional Biologist's Toolkit (2015)
The Dimensional Phenomenologist's (2016)

... ARTIST'S (2017)
... CRITIC'S (2018)
... EXCEPTIONIST'S (2019)
... UNIVERSALIST'S (2020)

... MATHEMATICIAN'S (2021)
... HISTORIAN'S (2022)
... POLITICS (2023)
... ECONOMICS(2024)

... PHYSICS (2025)
... POETICS(2026)
... TIME-TRAVEL (2027)
... IMMORTALITY (2028)

FICTION
    ONE-PAGE-CLASSICS
    LESSONS OF THE MASTER
    STORY OF MASTER WU
    DRAMATIS PERSONAE
    BANNED CLASSICS

Nathan Coppedge

BIO

On Nov. 9 - 10, 2013, Nathan finally built a device which he feels proves over-unity. Although the apparatus is not a perpetual motion machine, it shows a combination of properties which suggest excessive motion from rest, which is the criteria for over-unity.

Subsequently, Nathan has been experimenting with "Master Angle / Reverse Gravity," a special case that appears to permit limited upward movement through angled horizontal support.

Find the videos on Youtube by searching for "successful over-unity" and "master angle"

A video of Nathan's successful over-unity experiment is present at: http://www.academicroom.com/video/evidence-against-classical-model

AND PLEASE, I WELCOME REVIEWS OF MY BOOK ON AMAZON, BARNES & NOBLE, AND ELSEWHERE. EVEN YOUR OWN BLOG!